Peter Lax, Mathematician
An Illustrated Memoir

Peter Lax, Mathematician
An Illustrated Memoir

Reuben Hersh

Photographic research by Eda Gordon

AMERICAN MATHEMATICAL SOCIETY
Providence, Rhode Island

2010 *Mathematics Subject Classification.* Primary 01A60, 01A70, 01A72, 35A21, 35L05, 35L40, 76L05, 65M06, 35C08, 35Q53.

Cover photograph courtesy of the New York University Archives. All other photographic permissions and acknowledgments can be found on p. xi.

For additional information and updates on this book, visit
www.ams.org/bookpages/mbk-88

Library of Congress Cataloging-in-Publication Data
Hersh, Reuben, 1927–
 Peter Lax, mathematician : an illustrated memoir / Reuben Hersh ; photographic research by Eda Gordon.
  pages cm.
 Includes bibliographical references and index.
 ISBN 978-1-4704-1708-6 (alk. paper)
 1. Lax, Peter D. 2. Mathematicians—United States—Biography. I. Title.
QA29.L422H47 2015
510.92—dc23
2014036520

Copying and reprinting. Individual readers of this publication, and nonprofit libraries acting for them, are permitted to make fair use of the material, such as to copy select pages for use in teaching or research. Permission is granted to quote brief passages from this publication in reviews, provided the customary acknowledgment of the source is given.

Republication, systematic copying, or multiple reproduction of any material in this publication is permitted only under license from the American Mathematical Society. Permissions to reuse portions of AMS publication content are handled by Copyright Clearance Center's RightsLink® service. For more information, please visit: http://www.ams.org/rightslink.

Send requests for translation rights and licensed reprints to reprint-permission@ams.org.

Excluded from these provisions is material for which the author holds copyright. In such cases, requests for permission to reuse or reprint material should be addressed directly to the author(s). Copyright ownership is indicated on the copyright page, or on the lower right-hand corner of the first page of each article within proceedings volumes.

© 2015 Reuben Hersh. All rights reserved.
Printed in the United States of America.

∞ The paper used in this book is acid-free and falls within the guidelines
established to ensure permanence and durability.
Visit the AMS home page at http://www.ams.org/
10 9 8 7 6 5 4 3 2 1   20 19 18 17 16 15

To my honored and beloved children, Daniel and Eva, and grandchildren,
David, Jessica, and Zeev

Contents

Acknowledgments	ix
Permissions and Acknowledgments	xi
Introduction	xv
Chapter 1. A prodigy and his family have a narrow escape	1
The Holocaust	6
Lax family	6
Chapter 2. Manhattan, NY, and Manhattan Project. An army private among the "Martians"	17
Into the army	19
Manhattan Project. An army corporal among the Martians	20
NYU and the Courant Institute	33
Chapter 3. Family life: Son, husband, father, grandfather	35
Anneli Cahn	35
Jimmy	38
Johnny	42
Lori	45
Chapter 4. Early career	49
The advent of the computer	51
Becoming Peter Lax's student	55
Chapter 5. The famous CDC 6600 bomb-scare adventure	69
Chapter 6. Later career	79
Chapter 7. The queen of Norway	91
Entr'acte. Peter's stories	97
Chapter 8. Books	105
Chapter 9. Pure AND applied, not VERSUS applied	113
Computing	118
Partial differential equations and the real world, a beginner's primer	121
Thinking geometrically about dynamical systems: Phase space	125
Function spaces, functional analysis	126
Nonlinearity	129

Chapter 10.	Difference schemes. Shocks. Solitons. Scattering. Lax-Milgram. Pólya's curve, etc.	133
	Difference equations and computing	133
	Shock waves	135
	Solitons	142
	The nonlinear Schrödinger equation occurs in water waves	148
	Scattering theory	150
	Differentiability of the Pólya function	154
	Lax-Milgram lemma	156
	Algebraic hyperbolicity	156
Epilogue		159
Appendix 1.	Anneli Lax	161
Appendix 2.	John von Neumann: The early years, the years at Los Alamos, and the road to computing	167
Appendix 3.	The life of Richard Courant	175
Appendix 4.	Peter D. Lax curriculum vitae	187
Appendix 5.	The closed graph theorem	199
Appendix 6.	List of Peter Lax's doctoral students (from the Mathematics Genealogy Project)	203
John Lax		207
Appendix 7.	From *A Liberal Education*, by Abbott Gleason, pages 314–317 on John Lax	209
Appendix 8.	John Lax article on Chicago jazz musicians	213
References		229
Index		233

Acknowledgments

Profuse and ardent thanks are due to many. To Peter Lax himself, of course, first and foremost. And to my partner, Veronka. And to Eda Gordon for her brilliant, amazing contribution, finding beautiful, expressive photographs, and obtaining the needed permissions. Also to Lori Berkowitz Lax and to Anneli, Jimmy, Tommy and Timmy and Dee Dee, to Miriam Helbok and Eleanor Piel, to Carol Hutchins and Al Novikoff and Elena Anne Marchisotto, to Brittany Shields, to Nancy Cricco and the staff of the New York University Archive on the 10th floor of Bobst Memorial Library, to Professors Cathleen Morawetz and Louis Nirenberg and Joe Keller and Chandler Davis and Martin Davis and Fred Greenleaf and Emile Chi, and to Nick Unger and Dr. Robert Wolfe, to Dr. Marnie Greenwood and her partner, David. To Alex Chorin, Jeff Rauch, Mutiara Buys, and Stan Osher and Joel Smoller and Phil Colella. To Dean Porter Gerber, Roger Frye, Frank Wimberly, Edward Dunne. To Francoise Ulam. To Richard Courant.

I have used quotations from Peter's previous interviews: in *More Mathematical People*, by Jerry Alexanderson; *The Mathematical Intelligencer*, by Istvan Hargittai; and in the *Notices of the American Mathematical Society*, reprinted from the *European Mathematics Newsletter*, by Martin Raussen and Christian Skau. Especially valuable have been Peter's interviews with Phil Colella for the Oral History of Numerical Analysis of the Society for Industrial and Applied Mathematics. Colella was a student of Peter's student Alex Chorin. He became head of the Applied Numerical Algorithms Group of the National Energy Research Scientific Computing Center and is a member of the U.S. National Academy of Sciences; he is known for high-resolution adaptive mesh-refinement schemes for partial differential equations.

Permissions and Acknowledgments

The American Mathematical Society gratefully acknowledges the kindness of the following individuals and institutions in granting permission to reprint material in this volume:

Donald Albers: Photograph of Peter Lax and brother John, Photograph of Rózsa Péter, Photograph of Peter Lax and father Henry, Photograph of Peter Lax with older Jimmy.

Gerald L. Alexanderson: Photograph of George Pólya and Gábor Szegő, Berlin, 1924.

Archives of the Mathematisches Forschungsinstitut Oberwolfach (Gerd Fischer, Photographer): Photograph of Jürgen Moser.

Michael Atiyah: Photograph of Professors Michael Atiyah, Ralph Phillips, Donald Spencer.

Norman Bleistein: Photograph of Peter Lax and Burt Wendroff.

Esther Brass-Chorin: Photograph of Alexandre Chorin.

Mutiara Buys: Photograph of Mutiara Buys.

Civertan Grafikai Stúdió: Photograph of Heroes' Square, Budapest.

Jitze Couperus: Photograph of CDC 6600 computer with console.

George Csicsery: Photograph of Paul Erdős.

Dartmouth College Library, Rauner Special Collections: Photograph of John Kemeny.

Virginia Davis: Photograph of Martin Davis.

Sarah Gleason: Photograph of John Lax, in Abbott Gleason, *A Liberal Education*, Tide Pool Press, 2010, Appendix 7.

James Glimm: Photograph of James Glimm.

Eda Gordon: Photograph of Lax books.

Ruth Harten: Photograph of Amiram Harten, Palo Alto, CA, 1982.

Hearst Corporation: *Esquire Magazine* cover, September 1958 issue; Illustration from "Who Are America's 54 Outstanding Young Men...And Why," ibid., p. 33.

Heriot-Watt University: Photograph of Soliton Experiment, Union Canal, August 1995.

Geraldine Calisti Kaylor: Photograph of Jeffrey Rauch.

Joseph Keller: Photograph of Joseph Keller.

Henry Lax (Courtesy of James Lax): Photos of Anneli Lax and of Anneli and Peter Lax at the Courant Institute.

James Lax: Photos of Elizabeth Thomas, the Lax family nanny, with Jim Lax on his graduation from Columbia University, 1976; John and Marnie, Greece, 1978; Peter on the phone with the Norwegian Academy of Science and Letters learning of his award of the 2005 Abel Prize, with Lori by his side, 6:30 a.m.; Peter with family and guests at Prince Camp, Loon Lake; Lori and Peter with Pierre at his college graduation, 2013; Family Thanksgiving, 2013. L-R: Sabah D'Agrhi, Tommy, Timmy, Jim, Pierre, Peter, Nancy Foldi, Lori; Peter's grandsons Pierre, Tommy and Timmy at the Abel Prize ceremony.

Lori Berkowitz Lax: Photograph of Lori Berkowitz playing the viola.

Peter Lax: Photos: Peter told his friend and fellow mathematician Louis Nirenberg he was willing to explain the same thing more than once, but not more than ten times; Peter (left) and Enrico Fermi (right) on a hike in the mountains around Los Alamos with a colleague; Niels Bohr (left) and Richard Courant (right); Henry and Klara Lax; Henry Lax with sons Peter and John on an outing in the Grunewald, Hungary, 1931. Materal grandmother Ilka Kornfeld, née Nemenyi, watches in the background; John and Peter Lax at Prince Camp, Loon Lake; Still life painting by Peter Lax; Anneli and Peter at Courant Institute; Peter with his young son Johnny; Peter with his young son Jimmy; Grandma Klara reading to Jimmy, 1957; Grandpa Henry with Jimmy on his knee, 1957; Peter giving Johnny tennis advice; John and James; Anneli Cahn Lax; Anna Lesznai, artist and poet with whom Peter studied painting; Kurt Otto Freidrichs at work with Richard Courant, 1965; excerpt from Lax, John, "Chicago's Black Musicians in the Twenties: Portrait of an Era", Journal of Jazz Studies **1** (June 1974), 106-127. Courtesy of Peter Lax.

Los Alamos Historical Society Photo Archives (Jack Aeby, Photographer): Photograph of the Trinity Test.

Los Alamos National Laboratory Archives: Photograph of John von Neumann, Richard Feynman, Stanisław Ulam.

Louis Nirenberg: Photograph of Louis Nirenberg; Photograph of Peter Lax and Louis Nirenberg.

National Academy of Sciences, National Academies Press: Appendix 3: The Life of Richard Courant by Peter Lax. Reprinted with permission from *Biographical Memoirs*, Volume 82, 2003, pp. 79–95 by the National Academy of Sciences, Courtesy of the National Academies Press, Washington, D.C.

NASA/MSFC (Marshall Space Flight Center): Photograph of Schlieren Testing of SLS 70-Metric-Ton Configuration.

New York University Archives: Photographs of Jerry Berkowitz (with Harold Weitzner), Chapter 3; Jack Schwartz, Chapter 6; Paul Garabedian, Chapter 8; Molotov Cocktail, Chapter 5; UNIVAC Installation, Chapter 4; Cathleen Morawetz, Chapter 3; Peter Lax, Front Cover Photo; Courant Protest, May 1970; Chapter 5, "They Can't Kill Us All" Banner, Chapter 5; and Anneli Lax at the blackboard, Appendix 1.

Norwegian Academy of Science and Letters: Photograph of Peter Lax receiving Abel Prize, 2005; Photograph of KappAbel winners with Peter Lax, 2005.

Stanley Osher: Photograph of Stanley Osher.

Rényi Institute: Photograph of Dénes König.

Rutgers University, The State University of New Jersey (Nick Romanenko, Photographer): Photograph of Martin Kruskal.

Diego Torres Silvestre: Photograph of the Eldorado Building, 300 Central Park West, New York, NY, 2005.

Shelby White and Leon Levy Archives Center, Institute for Advanced Study (Alan Richards, Photographer): Photograph of John von Neumann and Robert Oppenheimer.

Society for Industrial and Applied Mathematics: "John von Neumann: The Early Years, the Years at Los Alamos, and the Road to Computing," *SIAM News*, Volume 38, Number 2, March 2005, pp. 9–10. Appendix 2.

Tide Pool Press, LLC: Excerpt from Abbott Gleason, *A Liberal Education*, Tide Pool Press, 2010, Appendix 7.

University of California, San Diego, Mandeville Special Collections, Leo Szilard Papers: Photograph of Leo Szilard.

University of Colorado Photolab Collection, Robert Richtmyer Archives, University of Colorado at Boulder: Photograph of Robert Richtmyer.

University of Texas at Austin, Dolph Briscoe Center for American History, Paul R. Halmos Photograph Collection: Photograph of Ralph Phillips.

Marina von Neumann Whitman: Photograph of John von Neumann, at about age 6.

John Wiley and Sons: Appendix 5: "The closed graph theorem", excerpt from Peter Lax, *Functional Analysis*, Copyright 2002 John Wiley & Sons. All rights reserved.

The following items are in the public domain:

Arrow Cross Leader Ferenc Szálasi, 1944; Photograph of von Neumann, Feynman, Ulam, attributed to Nicholas Metropolis, negative donated to LANL; Photograph of Stanisław Ulam.

Introduction

Who is Peter Lax? A famous mathematician. The major player for a half century in the dynamo of American applied mathematics. The last surviving member of the great wave of European mathematicians and physicists in the 1930s and 1940s who helped create the much-heralded "American century." (It turned out to be just a half-century.)

Back in my grad school days from 1957 to 1962, Professor Peter Lax was my mentor. Now we are both in our eighties and old friends. This book is more than just an account of a career in mathematics, with conjectures and theorems, professorships and prizes. It follows his life's arc, from childhood in the glamour and tragedy of Horthyite Budapest, to a lifetime of success in New York and Los Alamos, culminating in Oslo with acclamation from the queen of Norway.

His achievement forces one to see the contrasts and similarities between pure mathematics and applied mathematics. For the greatest mathematicians, the pure and the applied have for centuries easily fed and reinforced each other (Newton, Euler, Gauss, Riemann, Poincaré, Hilbert, von Neumann, Gelfand). Yet the value systems of the pure and the applied are opposed. Their prizewinners are almost two disjoint sets. Every day they compete for money from "granting institutions," including universities. They are a strange kind of twins, simultaneously incompatible and inseparable.

Peter is one of the greatest living applied mathematicians and one of the greatest living pure mathematicians. This book returns repeatedly to this duality.

In telling his life story, I cannot avoid honestly including my own political identity as a lifelong "lefty", "peacenik", and all-around dissenter. Peter and I came of age during the great war against Nazi fascism, and matured and lived through a half-century cold war of nuclear threat of mutually assured destruction. We shared many liberal values, and at the same time were separated by profound differences. Our disagreements have been very serious and difficult but always respectful, and in the end, perhaps, really not all that important as they seemed at the time.

Politics entered Peter's life story repeatedly, first in the form of murderous Hungarian and Nazi anti-Semitism, then in conflicts over accepting refugees and immigrants into the U.S. In the famous Manhattan Project, where a new kind of bomb was being invented and manufactured, Peter was initiated into serious applied mathematics. During the Vietnam War, in a protest at NYU, Peter played a surprising and courageous role.

I, growing up as a New York Jew in the Depression decades, took left-wing sympathies for granted. To him, growing up in the aftermath of a disastrous failed

Hungarian Bolshevik revolution and knowing the dire fate of Hungarian Jewish Bolsheviks who took refuge in Stalin's Moscow, fear and hatred of Soviet Communism were taken for granted.

In writing this book I first discovered that fifty years earlier, at a time when I as his student felt a great ideological gap between us, Peter was quietly giving essential material support to my school friend Chandler Davis, who had been blacklisted by U.S. academia after serving prison time for contempt of the U.S. House of Representatives Committee on Un-American Activities.

Lax is an ex-prodigy who more than fulfilled his promise. To a distinguished family heritage, Peter brought a brilliant talent, which was recognized early and carefully nurtured both in Budapest and in New York.

Starting in Budapest, we progress through his arrival in New York, his service as a corporal on the Manhattan Project in Los Alamos, his family life, his career, his famous adventure with the bomb scare at NYU in 1970, and on through his writings, his students, and his prizes. I present his remarkable discoveries and creations without requiring advanced technical preparation. On certain pages, there is a brief reversion to the pedagogic or textbook tone, but in the main I present his mathematics as a narrative.

He was recognized early as a likely future master mathematician. In 1941, almost at the last minute, he escaped with his family to comfort and privilege in Manhattan. When they arrived in New York, they brought with them letters from his mentors in Budapest to two Hungarian mathematicians already in the U.S.: John von Neumann in New York and Gabor Szegö at Stanford University in California, whose wife was Peter's mother's cousin.

Von Neumann suggested Peter work with his collaborator Francis Murray at Columbia University. But Szegö said that Richard Courant at New York University was "very good with young people." Peter later said that Szegö's advice was the best possible. At sixteen he became a student at NYU, starting to grow into his role as a superstar there. He remained there as a professor through his whole career. NYU's graduate math program in time would be renamed the Courant Institute.

From his first days in the U.S., he was guided, supported, and inspired by Richard Courant and John von Neumann. These two émigrés were different in many ways, but in certain respects they were similar. They both combined deep devotion and commitment to mathematical values (truth, profundity, elegance) with high achievement in the American stratosphere of money and politics.

Von Neumann, one of the supreme mathematical intellects of the twentieth century, would be Peter's lifelong model of excellence in mathematics.

Included in this book are biographies that Peter wrote of Courant and von Neumann, so it is unnecessary for me to say much about them here.

Courant is unique in the history of mathematics, the leader of two great mathematical centers—first in Göttingen, Germany, in the 1920s, and then in New York from the 1930s for the next half century. Courant is the author of three great books. The first, affectionately called just "Courant-Hilbert" or, more formally, *Methods of Mathematical Physics*, is based on David Hilbert's lectures and became an indispensable handbook for the physicists who created quantum mechanics in the 1920s. Then there is his textbook on differential and integral calculus, which is still the very best of all. And there is *What Is Mathematics?*, a delectably inviting work of art that opens modern mathematics to the eyes of any literate intelligent reader. (The co-author was the future famous statistician Herbert Robbins.)

Von Neumann was legendary for sheer mathematical power. Lightning fast computation, encyclopedic memory, uncanny ability to connect seemingly remote mathematical concepts. He first became known for contributions to the "foundations of mathematics", axiomatic set theory. Then with Oskar Morgenstern he invented game theory, the mathematics which is still today the favorite plaything of high-powered strategists. Then he used a new infinite-dimensional function space, which he named "Hilbert Space," to unify and establish rigorously the rival quantum mechanics of Heisenberg and Schrödinger. After that he deeply explored the theory of linear operators, introducing "pointless spaces" and continuously varying dimension.

By the time Peter met him in wartime in the 1940s, von Neumann had blossomed as an applied mathematician, a famously powerful one in the Manhattan Project at Los Alamos, New Mexico. There he was joined by eighteen-year-old Corporal Peter Lax. Von Neumann was already one of the inventors of modern digital computers. One of the first, called the MANIAC, was also sometimes the "Johnniac". He was fascinated by a difficult mathematical-physical problem: shock waves, such as the implosion wave he used to detonate a plutonium bomb or the tremendous pressure wave that springs out of an atomic explosion. These are interesting physical phenomena. Mathematically speaking, they are discontinuous solutions of certain nonlinear partial differential equations. Predicting where and when the discontinuity arises and how it propagates remains to this day a deep, difficult problem.

Peter worked on shock waves alongside John von Neumann and continued and amplified von Neumann's work after von Neumann's premature death.

Von Neumann was famously influential at the highest levels of U.S. decision making. When asked about this during a television interview in their native Hungary, his friend Eugene Wigner explained that "after von Neumann analyzed a problem, it was clear what had to be done."

Courant also was an adviser and consultant in high circles, but most important was his great skill and success in promoting, establishing, and enlarging the institution that later would bear his name. It was no simple or straightforward matter for NYU to become the principal academic center of applied mathematics in the U.S. To the intense competition for federal support Courant brought persistence, guile, and skill at winning the friendship of powerful people.

One of the pleasures of my research for this book was finding in the NYU archives the recommendations that Courant wrote for Peter Lax, including his recommendation for Peter's election to the National Academy of Sciences.

Peter Lax says that he likes to pick out a striking phenomenon, such as shock waves, and analyze it mathematically. While such work is motivated by physics and produces information valuable to physicists, the analysis itself, as done by Lax, is pure mathematics. It is precisely stated as a mathematical problem, and the claimed result is rigorously proved. Ordinarily, in applied mathematics, in search of a useful result, one may take a shortcut, justified more by hope or by intuition than by rigorous logic. Peter's work on shock waves, on scattering of light and sound waves, and on the Korteweg-de Vries equation of fluid dynamics is not like that. It is totally and strictly rigorous.

Is Peter Lax a pure mathematician or an applied mathematician? He sees no difference between the two. By choosing problems from physics and solving them as problems in pure analysis, Peter unites the pure and applied seamlessly and inseparably.

In the larger mathematical world, the distinction between the pure and the applied is clear. Applied mathematics values utility and relevance, giving the customer the information needed to get the job done. Rigor and elegance can come later, if at all.

Applied mathematics is likely to be found in an industrial laboratory or in university departments of computing or engineering or applied mathematics. There are journals of applied math, and there are journals of pure math. Some articles could appear in either place, but many could appear only in one or the other. In the pecking order of academic prestige, pure mathematics has a lofty position, looking down from above at any sort of application. But in the practical worlds of money and power, applied mathematics is recognized and important, while the pure in its ivory tower at best receives faint praise.

This de facto separation into two social/academic worlds is mocked by the many very important scientific linkages between them. Applied mathematics often provides great inspiration for work in pure mathematics. Pure mathematics constantly provides essential tools for applied mathematics. At his institute, Courant recognized no separation or distinction between the two. Courant shared this philosophy with his mentor, David Hilbert.

Peter Lax's career is a singular exception to the usual mutual disrespect between these two inseparable and incompatible twins, the pure and the applied. In phenomena of visible and palpable physics, he found deep mathematical problems whose statement and whose final result or solution can be understood by any interested reader. For other prizewinning mathematicians of recent decades, even the titles of their epochal discoveries are not decipherable by the ordinary practicing mathematician.

Peter Lax is the rare mathematician who is completely at home and masterful in the abstract axiomatic style while still rooted in and always inspired by concrete physical phenomena, such as shock waves, for example.

Students sometimes come to the university anticipated and well known, having shown brilliant promise to qualified judges. Such great promise is not always followed by the hoped-for great performance. The aspiring mathematical genius usually has years of struggle ahead. First, struggle to get into the best graduate program, then struggle to beat out other aspiring grad students for the best thesis adviser, then struggle to make enough of a dent in the right problem to impress the adviser and the broader research community, then struggle for tenure at a top-notch math department, then struggle to make more great discoveries—arduous years to make it to the top and stay there.

Peter is one of the few who rose to the top and stayed there, from childhood through the rest of his life. Great promise in childhood was fulfilled by great achievement as a professional. Among the international mathematical giants of our time, Peter Lax is almost unique. He has one foot in each of two camps: one side is the deeply abstract and theoretical, the other side is the computational, practical, scientifically useful.

Peter told his friend and fellow mathematician Louis Nirenberg he was willing to explain the same thing more than once, but not more than ten times.

Louis Nirenberg is one of the most honored living mathematicians, the uncontested world expert on inequalities and estimates. He says,

> Peter always knew a lot more math than I did. When he was explaining something to me he sometimes said he was willing to explain the same thing more than once but not more than 10 times. Peter has been like a brother to me.

Peter's life was not free from misfortune or suffering. With his family he had to flee the murderous Nazi persecution. Later he suffered very painful losses of his closest loved ones. But he doesn't show scars or wounds. He's remarkable for kindness and gentleness.

I chatted with Peter's grown-up grandsons, Tommy and Timmy, and asked Tommy, "What do you think of Peter as a grandfather?"

He answered, "My earliest memory of him is that Peter was the only person that could really get me out of bed."

"How did he do it?"

"He would sit right next to me and sing Irving Berlin's 'You gotta get up, you gotta get up, you gotta get up in the morning!' with a Hungarian accent."

My friend Mutiara Buys, a mathematician whose Ph.D. is from NYU, writes, "For two years I made sure I attended at least one of Peter's classes, and continued sitting in his classes and seminars after orals. I was attracted to his extremely lucid and clear mind, as well as his charming and witty personality.

"The proverbial prof just goes on ad infinitum writing math computations on the blackboard. Little time is devoted to thinking about what a formula means. Peter Lax devotes considerable attention to speaking about and conceptualizing a computation, so valuable insights are imparted to the student.

Mutiara Buys in front of Bobst Library, NYU, 1980

"One day Peter was stuck in one of his computations and was rather annoyed with himself. He stared at the blackboard for a long stretch. All of a sudden he hit his head dramatically and exuberantly shouted, "Oh my God, I could have had a V-8!" (famous commercial for the vegetable juice V-8 of a guy hitting his head; I hope you know it). He energetically and happily erased the board and provided a succinct, elegant proof while apologizing for not having had a good proof to start with. I don't think I have ever not laughed in his class; he is an engaging and fun lecturer. Outgoing, amiable, and fun-loving.

"At a wedding attended by many shy people, Peter was on the dance floor showing off his dance steps and egging people on to participate.

"At my thesis defense he was dozing off. To get his attention, I had to loudly mention the Lax Pair, at which he immediately perked up and listened attentively."

This last experience of Mutiara's was not unique. Less famous than his mathematical brilliance and his personal kindness and friendliness is Peter's narcolepsy (falling asleep inappropriately). As a fellow sufferer from this disability, I know that it is a misfortune, amusing though it is to onlookers.

At a time during my student days when I didn't have an office, Peter offered me a desk in his office. He was visited by Seymour Parter, a respected numerical analyst at the University of Wisconsin, who was eager to talk about his great new idea (to use graph theory to study matrices for numerical calculation). Peter's head would nod off as he lapsed into slumber, but Seymour just shouted, "Wake up! Wake up! Peter, wake up!" Peter would shake his head, open his eyes, smile slightly, and resume listening to Seymour's lecture.

Peter's telephone was constantly ringing. He was always willing to interrupt his work and help the caller with his problem. Once I actually was able to help Peter Lax. He received a phone call from the ground floor, "Mr. So-and-so is here to take his real variables exam." In our course on real variables, Peter had made the final exam optional, and now here was a student actually asking to take a final! The last thing Peter wanted was an interruption to write a real variables exam. He moaned, "What can I do?" "It's easy," I answered. "So-and-so doesn't want to settle for a B, he wants an A. So just give him an A." "Brilliant!" cried Peter. He picked up the phone and told the waiting receptionist, "Tell Mr. So-and-so that his grade for the course is A." Problem solved.

Alex Chorin, a professor at Berkeley, is an original, influential contributor to fluid dynamics, but as a graduate student, he had to struggle. "It is very clear to me that Peter had some idea that I was having difficulties not necessarily related to mathematics and was very patient with my absences and my not always turning up when summoned. I was grateful then for Peter's patience and humanity, and am even more grateful to him now that I have had grad students of my own. Without his kindness I doubt I would have received a Ph.D. It is not easy to see the need for kindness when the usual view of one's duty is to insist on steady progress."

CHAPTER 1

A prodigy and his family have a narrow escape

Peter Lax was born in Budapest, Hungary, and lived there for his first fifteen years, when with his family he escaped to the United States in time to avoid the Holocaust. He has lived in the U.S. since 1941 and has long been thoroughly Americanized, but still cherishes his heritage of Hungarian language and culture. Therefore, it is necessary to start his biography with essential background: the strange and fascinating history of the Magyars.

The Hungarian nation and its language are often called "Magyar". The Hungarians or Magyars entered Central Europe from Asia in the ninth century. Their language has always set them apart, for it is neither Slavic like Czech or Polish nor Germanic nor Romance. In the middle of Europe, with neighbors speaking German or Romance or Slavic languages, Hungarians speak a "Finno-Ugric" tongue distantly connected to Finnish, Estonian, and to small populations in Siberia. There are hardly any commonplace Hungarian words whose meaning can be guessed even if you know English, French, German, and Latin. (I once attempted attending a class in Hungarian for beginners, but I was soon defeated.) An educated Hungarian must be bilingual, for rare is the non-Hungarian who will speak Hungarian.

Hungary was converted to Christianity early. The majority of its population are Roman Catholic, and the cardinal and his bishops play a powerful and malignant role in Hungarian politics. Hungary was affected by the Protestant Reformation, and a substantial minority of the population are Lutherans or Calvinists. There are also Jews.

Like its neighbors Romania, Bulgaria, Serbia, Croatia, Slovenia, Albania, and Greece, Hungary for a time fell under the Ottoman or Turkish Empire. For one hundred fifty years the central part of Hungary was ruled by the Ottoman Turks. The Turks had conquered the Byzantine Empire, which was the relic of the eastern half of the Roman Empire. In a famous battle in 1683, the Holy Roman Empire, whose capital was Vienna, turned back the Ottoman invaders. As a consequence, Hungary became subject to the emperor of Austria. Hungary was then three times as big as today. It included not just Magyars and Jews but also Romanians, Serbians, Croatians, Rusyns, Germans, and Slovaks. It had access to the sea by an Adriatic coastline and many large productive plains where grain grew and cattle were raised. The land was divided among large estates owned by Magyar noblemen, who lived richly from the labor of oppressed peasants. Like noble landowners in other countries, these Magyar rulers loved hunting deer and boar and being soldiers. Rarely were they scholarly or businesslike. One famous noble family, the Esterházys, were patrons of Joseph Haydn, the friend of Wolfgang Amadeus Mozart and teacher of Ludwig von Beethoven. The Esterházys could easily afford their own private orchestra and composer. In western Europe, Hungary was known as a wild, romantic, less civilized country, with incredibly rich noblemen. The operetta *Die*

Fledermaus by Johann Strauss Jr. centers on a fancy-dress ball given by a fantastically rich Hungarian count who lives in Vienna on the income from his huge Hungarian estate.

In old Hungary, Jews were peddlers, tavern keepers, teamsters, tailors, shoemakers. Some rabbis, of course, and some estate managers, bookkeepers, moneylenders. They lived in their city districts ("ghettos"), with their ancient religion. Noncitizens, without civic rights. At the end of the eighteenth century they were by and large emancipated by the liberal emperor Joseph II.

In 1848 Europe was shaken by revolution. Across the continent, people struggled to break from shackles that had been imposed by Russia, Austria, and Prussia after the defeat of Napoleon. Hungary tried to escape its domination by Austria's Habsburg emperor, but its war of independence was lost when Russia came to aid Austria. Then in 1867, after losing a war with Prussia and while struggling against the rising nationalism of his Balkan subjects, the Habsburg emperor in Vienna made a deal with the Hungarian aristocrats and upper classes. In return for their support of his rule, he promoted the Magyars to partners. His empire became an Austro-Hungarian Empire, with Budapest a second capital after Vienna. The empire became "the double crown". Hungary was allowed self-government, except for foreign policy and the military. The Magyars became Austria's agents, suppressing oppressed nationalities between the Carpathian Mountains and the Adriatic Sea. This arrangement was ever after called the *Ausgleich* (the Compromise). And the Jews became allies of the Austrians and the Hungarians.

The economic development of the mid-century and the *Ausgleich* opened the door for Hungary to enter the modern age of industry and commerce. But to do so required entrepreneurs—energetic people who knew how to innovate, how to modernize, how to take risks. Who was capable of constructing a modern metropolis for a still agrarian country? Most Magyar nobility had neither inclination nor preparation for commerce. They were known to take pride in letting hirelings do the work. The Magyar peasant was not about to found a business or enter a profession. Together with some German immigrants and the more enlightened section of the Magyars, it was left to the Hungarian Jews to make Budapest the exciting, dynamic capital of modern Hungary. They were granted citizenship, they could move to Budapest, they could start businesses, they could enter the professions! The Hungarian Jews responded eagerly. They became passionate Hungarian nationalists, deeply loyal to the Magyar nation and to its Hungarian king/Austrian emperor. They proudly declared themselves Hungarians first, Jews second.

By 1900, Budapest was comparable to New York—a busy center of commerce, with vibrant journalism, theater, art and music. There was a brilliant literary flowering. It had the largest stock exchange in Europe and the most grandiose parliament building in the world.

"Booming Budapest of 1903, into which Johnny von Neumann was born, was about to produce one of the most glittering single generations of scientists, writers, artists, musicians, and expatriate millionaires to come from one small community since the city-states of the Italian Renaissance. In much of 1867–1913, Budapest sped forward economically faster than any other city in Europe, and with the delights that a self-reliant plutocracy (rather than self-questioning democracy) temporarily brings. Budapest surfed into the twentieth century on a wave of music and operetta down the blue Danube, as an industrializing city that 'still smelled

Heroes' Square, Budapest–a World Heritage Site to the 1000-year history of Hungary

of violets in the spring,' pulsing with mental vigor in its six hundred cafes and its brilliant elitist schools." (Norman Macrae, quoted by George Marx, page 267.)

"Budapest rivaled Paris and Vienna in first-class hotels, garden restaurants, and late-night cafes, which were hothouses of illicit trading, adultery, puns, gossip and poetry, the meeting places for the intellectuals and those opposed to oppression. It was an Old World city, its women praised for their beauty and its men for their chivalry." (Paul Hoffman, *The Man Who Loved Only Numbers*, quoted on page 280 of Marx.)

Jews were prominent. Many took Magyar names. Among famous mathematicians, Weisz became Fejér, Politzer became Peter. Jewish bankers and industrialists became nobles and barons, with hereditary titles. The famous mathematicians John von Neumann and Theodore von Kármán inherited their honorific "von" from their ennobled fathers. Theodor's father, Mor, was instructor of the crown prince and a reformer of education.

As in Paris, London, and Berlin, the period before World War I in Budapest was later remembered nostalgically as a lost time of bourgeois prosperity, stability, and optimism. This prosperous comfort was of course based on the misery of an industrial proletariat and a downtrodden peasantry. But for the striving middle class, there was good music and literature and a pleasant culture of theaters, newspapers, concerts, and coffee houses.

One famous Jewish family were the Kornfelds, whose dynasty was founded by the banker Zsigmund (1852–1909). He was the grandson of Aharon Kornfeld, the last great head of a yeshiva in Bohemia. He came as a youth from Vienna in 1878 to run the Rothschild banking empire's branch in Budapest. Only twenty-six years old, he was chosen by Albert Rothschild to be director of the Hungarian General Credit Bank in Budapest.

His bank founded and reorganized industrial companies, including a mining company in Bosnia and an oil refinery in Fiume (now Rijeka, Croatia). He participated in founding the Budapest-Pécs Railway, the Electric and Transportation Share Company, and the Hungarian River Navigation and Maritime Share Company, of which he became president. In 1892 he was instrumental in a currency reform. In 1900 he became the managing director of his bank, and in 1905 the president. In 1899 he was elected president of the Budapest Commodity and Stock Exchange, where he made Hungarian replace German as its official language. In 1909 the Emperor Franz Joseph gave him the title of baron.

The First World War resulted in catastrophic losses for Hungary and Hungarian Jews. The Austro-Hungarian or Habsburg Empire, along with Prussian Germany and Ottoman Turkey, was defeated by England, France, and the U.S. (The other "ally," the czar of Russia, had been knocked out of the war by the Bolshevik Revolution.) The Treaty of Trianon, imposed on Hungary by the victorious Allies, liberated the formerly oppressed nationalities of the Austro-Hungarian Empire by creating two new states, Czechoslovakia and Yugoslavia, and giving the large province of Transylvania to Romania. Hungary was reduced in size to one third, lost much productive farmland, and was landlocked, cut off from the Adriatic Sea. Many Magyars became minorities living in Czechoslovakia, Yugoslavia, or Romania. For decades afterward, Hungarian schoolchildren started each day by chanting, "Nem, nem, soha!" (No, no, never!) (Never accept the loss of Hungarian territory, the Treaty of Trianon!)

The destruction of the Austro-Hungarian Empire and the punitive Treaty of Trianon were followed by two more disasters. A Hungarian prisoner-of-war in Siberia named Béla Kun became a follower of Lenin and returned to Budapest to lead a four-month Bolshevik revolution. Homes and businesses of the rich and prosperous were expropriated. New radical officials took government office. Kun himself and many of his prominent followers were Jews (although of course nonobservant and not connected to the official Jewish community of Budapest). Many of the homes and businesses that they expropriated had been owned by prosperous Jews. This experience left long and bitter hatred of communism or socialism among many middle-class Budapest Jews.

After four months, the Bolshevik regime in Budapest was destroyed by an invading army from Romania. Hungary was restored as a "kingdom" without a king. A Magyar admiral named Miklós Horthy was appointed "Regent" and would rule Hungary for three decades. His regime was destroyed by the Nazis, and then by the Soviet army, at the end of the Second World War.

Horthy's regime was a nondemocratic, authoritarian system. It started out with two years of White Terror, taking merciless revenge on the Bolsheviks and on their Jewish supporters and sympathizers. Béla Kun escaped to Moscow as a leading "functionary" in the Communist International, to be executed years later in one of Josef Stalin's fits of murderous paranoia. (Kun's grandson Miklós returned from Russia to Hungary and is a historian there, a specialist in "Kremlinology"—the private life of Josef Stalin.)

The Horthy regime was unremittingly plagued by the "Jewish question". Throughout the 1920s and 1930s, anti-Semitic policies and political parties grew more prominent. In 1929 came the worldwide stock market crash and subsequent Great Depression, with a terrible crisis of unemployment and deprivation. The

Arrow Cross Leader Ferenc Szálasi, 1944

scapegoating of Jews intensified. As the 1930s and 1940s advanced, so did horrifying fascism and anti-Semitism in both Hitler's Germany and Horthy's Hungary. In Hungary arose a Nazi-type anti-Semitic party, "Arrow Cross". Their disgusting and frightening signs and slogans multiplied, their power and influence grew.

Horthy imposed quotas on Jewish professors and students at the university and restricted Jewish participation in the professions and the civil service, but he continued faithfully to respect private property and capital. Major armaments industries continued to be owned and run by Jews. Major publishing companies, theaters, and coffee shops were still owned and patronized by Jews. The Jewish population of Budapest was around 30 percent. The prosperity, glamour, and brilliance of Budapest life were to a great extent created by Jews, who were prominent in banking, industry, journalism, law, and medicine. Until the Nazi clamp-down in the last months of the Second Great War, a major, almost dominating, role in the city continued to be played by Jewish professionals and capitalists.

Around 1936 the desire to recover its lost territories pushed Hungary towards the German-Italian axis. Anti-Semitism was already widespread in Hungary, and closer ties with the Nazis made the situation much worse. Anti-Jewish laws were passed, partly to please Hitler. The Catholic hierarchy and the Magyar nobility had formerly accepted the "over-talented, over-prosperous" Jews as allies against the other nationalities. But now Jews came to be seen as outsiders who had somehow grabbed more than their fair share. No matter how thoroughly the Jews Magyarized themselves—even becoming champion fencers and water polo players—they were seen as just not really Magyars.

In 1938 and 1940 Hungary seemed to be achieving its territorial objectives when the Vienna Awards returned some of its lands. In 1939 World War II began, but Hungary stayed out of it until 1941, when it joined the war against Russia on the side of Germany. By then, Hitler had ghettoized and imprisoned the whole German Jewish population, conquered Czechoslovakia, split Poland with Josef Stalin, and overrun Belgium, Holland, Denmark, Norway, and France.

To these events, the dominant leadership of the Hungarian Jews and the bulk of their population responded by declaring themselves more and more Hungarian. (And of course a substantial portion of Hungarian Jews took comfort and refuge in their religion.) Zionist and socialist alternatives did not become influential. After all, they were faithful, loyal Hungarians! They did not recognize that their prewar status as needed defenders of the status quo had disappeared. There still remained in Budapest a large, vibrant Jewish community that produced famous writers, musicians, and scientists.

A very prominent Jewish publisher named Vészi was a friend of Peter's father. As a child Peter played with cousins on a large estate belonging to Vészi. When he revisited it decades later, it had fallen into ruin.

The Holocaust

On March 19, 1944, Berlin got wind of cease-fire talks between Hungary and the Allies. German troops occupied Hungary. Accompanying the German army was a special unit (*Sondereinsatzkommando* (SEK)), commanded by SS Lieutenant Colonel Adolf Eichmann, with orders to "dejewify" the country.

The SEK was a force of about one hundred people, including twenty officers, and drivers, guards, and secretaries. Without help, these one hundred Germans could not annihilate 770,000 Jews scattered all over Hungary. Regent Miklós Horthy appointed a new government to serve the Nazis, led by Döme Sztójay. Between mid-April and late May, in the largest deportation ever achieved during the Holocaust, practically the entire Jewish population of the Hungarian countryside was ghettoized. Between May 15 and July 9, over 437,000 of them were delivered to Auschwitz.

By early July 1944 only Budapest Jews and those serving in labor service units were still in Hungary. Then the deteriorating war situation and the spreading news of mass extermination convinced Horthy to try to stop the deportations and escape from the German alliance. So on October 15 the Germans forced Horthy to resign and gave power to Ferenc Szálasi of the Arrow Cross movement. In November and December 50,000 more Jews were taken to Germany, most of them driven there on foot. The Jews who remained in Budapest were locked into two ghettos where Arrow Cross militia murdered thousands of them.

Between 1941 and 1945 more than half a million Hungarian Jewish men, women, and children were destroyed on the streets of Budapest, in countryside ghettos, behind barbed wire in German concentration camps, and in gas chambers.

Lax family

Before all these calamities, Peter Lax's family lived a happy and prosperous life in Budapest. (The name "Lax" is the same word as "lox", the familiar New York snack. It just means "salmon".) His father, Henrik, was an internist, and his mother, Klara, after training as a pediatrician switched to clinical pathology and ran Henrik's laboratory. Henrik and Klara met in medical school. Peter thinks Klara may have been one of the first women admitted to medical school in Hungary. Henrik was chief of medicine at the Jewish Hospital in Budapest and famous as a brilliant diagnostician. His patients from high society included the playwright Ferenc Molnar, the composer Béla Bartok, the film magnate Sir Alexander Korda, and the Swedish actress Greta Garbo. Molnar would later serve as the Lax family's

Peter and his brother John in their childhood in Hungary

sponsor in the U.S. (Someone had to promise to support them financially if ever they became charges on government charity!) Korda expressed his appreciation of Henrik's treatment by a gift of the painting *Waterloo Bridge* by the French impressionist Monet. (Worth $10,000 in 1952, it ultimately turned out to have a truly incredible cash value.)

Henrik told Peter about one of his most impressive cases. A diabetic patient was approaching death, even though the insulin Henrik prescribed should have maintained her. Henrik brilliantly solved the mystery: her husband was diluting the dose in order to murder her! However, Doctor Lax was not rewarded by the patient's gratitude. On the contrary, she reconciled with her would-be murderer and changed doctors!

Peter was born on May 1, 1926. He had an older brother, John, as well as a cook and a *fraulein* or nanny. It was customary to hire a young Austrian woman for child care, benefitting the child by exposure to the German language. Peter says, "My brother and I had a series of nannies; we were rather undisciplined, so they quit, except the last one, who stayed for five years. She survived the war and afterwards emigrated to Australia. We kept in touch with her."

The family life was taken up with socializing, listening to fine music, and playing tennis. They were Jewish by descent but not at all observant. Peter's paternal grandmother in her youth had been a suffragette!

Peter early displayed a talent for numbers. His mother always liked mathematics, and her cousin was married to a famous young mathematician, Gabor Szegö. Mathematicians refer to the 1920s and 1930s in Hungary as "the Hungarian miracle". There were the exceptionally brilliant John von Neumann and the brothers Frigyes and Marcel Riesz, the collaborators George Pólya and Gabor Szegö, Leopold Fejér, Cornelius Lanczos, Bela Szőkefalvi-Nagy, Arthur Erdelyi, Paul Erdős, Dénes König, Rózsa Péter, Michael Fekete, Paul Turan. Of this list, all but Szőkefalvi-Nagy were Jewish!

Janos Bolyai, one of the creators of non-Euclidean geometry in the early nineteenth century, was a hero of Hungarian culture, even though in his own lifetime he was totally ignored. Two educational institutions helped greatly in fostering young Hungarian mathematicians. A high school math newspaper, *The Mathematics Journal for Secondary Schools* (*Középiskolai Matematikai Lapok*), founded in 1894 by Daniel Arany, published mathematical problems and solutions. The possibility of getting one's name in the paper fostered a great spirit of friendly competition among young problem solvers.

And there was a problem-solving contest for high school graduates, called the Eötvös Competition, in honor of Baron Lorand Eötvös. He was a leading physicist, internationally recognized for precise measurements of the gravitational force. Many future Hungarian mathematicians and physicists were prizewinners in the Eötvös Competition.

Peter thinks some credit for the Hungarian Miracle should go to an earlier Hungarian mathematician, Julius König. He was a student of Leopold Kronecker, made an important contribution to Cantor's set theory, and was influential in nurturing mathematics in Hungary. His son, Dénes, is considered the father of modern graph theory. Dénes befriended and encouraged Peter and sent a letter of support to John von Neumann.

Peter likes to tell how Leopold (Lipot) Fejér, the first Jew proposed for a professorship at Budapest University, won his appointment there. "At that time there was a very distinguished Jesuit theologian, Ignatius Fejér, in the Faculty of Theology. One of the opponents, who knew full well that Lipot Fejér's original name had been Weisz, asked pointedly: 'This Professor Leopold Fejér that you are proposing, is he related to our distinguished colleague Father Ignatius Fejér?'

Rózsa Péter (1905–1977)

And Lorand Eötvös, the great physicist who was pushing the appointment, replied without batting an eye: 'Illegitimate son.' That put an end to it."

Peter's uncle Albert Korodi, his mother's younger brother, an engineer by profession, was "a terrific mathematician." (He had changed his name from Kornfeld.) At eighteen he won the mathematics prize in the Eötvös competition. That year Leo Szilard won the physics prize, coming in second after Korodi in mathematics. Szilard and Korodi became very good friends.

Szilard is famous for first conceiving of a chain reaction of uranium fission. He made important inventions, including patent applications for a linear accelerator in 1928 and a cyclotron in 1929. He worked with Albert Einstein to develop a refrigerator that had no moving parts. It wasn't commercially successful, but it turned out to be useful in experimental physics. Peter's uncle Albert Korodi did the engineering design for the Einstein-Szilard refrigerator.

According to the Hungarian physicist George Marx, who heard the story from Korodi (*The Voice of the Martians*, page 213, Akademai Kiado, Budapest, 2001), "Szilard's most famous invention was that a household refrigerator would last longer

Dénes König (1884-1944)

if the compressor pumps had no rotating solid parts subject to abrasion. He wished to use the Lorentz force, exerted by a static magnetic field, on direct current flowing through mercury, to drive the liquid metal around. He asked Albert Korodi, his friend since their simultaneous winning of the Eötvös Competitions, to develop this idea. Korodi was studying engineering in Berlin and made detailed plans and calculations, with the conclusion that the efficiency of the magnetic compressor would be very low, due to the modest electric conductivity of mercury, compared to that of copper. A week later, Szilard turned up again, suggesting the use of a eutectic mixture of sodium and potassium, liquid at room temperature and making a good conductor. Practical studies by Korodi showed that sodium and potassium were too abrasive, attacking the insulation of the wires conducting the electric current in the airtight container. Szilard complained to Einstein about his difficulties. After a few minutes of thinking, Einstein proposed an arrangement in which the current was closed through an air gap within the container, thus the corrosion of the insulator could not endanger the vacuum-tight enclosure of the aggressive liquid metal. Einstein agreed to patent the magnetic compressor under the names of Einstein and

Szilard, possibly helping to improve Szilard's financial position. Szilard convinced *Allgemeine Elektrizitatsgesellschaft* to build a prototype of the magnetic refrigerator. The company hired Albert Korodi to construct the machine. The practical efficiency of the fridge turned out to be still too low, therefore the system did not catch on. But liquid sodium driven by a magnetic pump is used nowadays to cool high temperature reactors, especially the breeders."

In 1939, living in the U.S., Szilard learned that physicists in Germany had achieved nuclear fission. Terrified at the possibility of a Nazi atomic bomb, he used his friendship with Einstein to alert President Roosevelt to the importance of an atomic bomb. As a remote consequence, five years later Peter Lax, then a private in the U.S. Army, would find himself assigned to something called the Manhattan Project in Los Alamos, New Mexico.

Peter learned a lot from his uncle Albert. "When I was about twelve I learned from him that, by the distributive law, -1 times -1 equals $+1$. I thought that was great. I remember my mother describing what a shock it was, when her younger brother came along, to realize on what a higher level mathematics can be understood. She remained quite interested in mathematics, and my father respected it from afar.

"My uncle Victor Farkashazi was a pediatrician, and he had a deep interest in mathematics. I still remember two wonderful problems he posed. Here is the first:

"A gold merchant wants to buy gold dust from gold miners in multiples of 1 ounce, up to 40 ounces. He has an old-fashioned kitchen balance to weigh the gold dust. Question: how many weights does he need? (Today the story would be told about a cocaine merchant.)

"And here is the second: A spider is sitting on the ceiling of a room, 1 foot from one wall, and 5 feet from the other wall. Its nest is located on the other wall, 1 foot from the first wall and 6 feet from the ceiling. She needs to get to her nest as fast as possible (I forget why). Question: what path does the spider take?

"What I like about these problems is that they illustrate basic mathematical principles."

Peter became very interested in math when he got to the Gymnasium, the famous *Minta* (Model School). It was founded as a teacher-training school by Mor Kármán at the suggestion of the minister of culture, Jozsef Eötvös, father of the physicist Lorand Eötvös. Other graduates of the *Minta* include the great aerodynamicist Theodore von Kármán, who was Mor Kármán's son, the chemist-philosopher Michael Polányi, and the H-bomb physicist Edward Teller.

Peter's parents provided him with a private tutor in mathematics. There were in Budapest quite a number of unemployed brilliant young mathematicians, mostly Jewish. Agnes Berger, a university student whose parents were physicians and close friends of Peter's parents, made the perfect suggestion for Peter's tutor: Rózsa Péter. Ultimately, Dr. Péter became famous for two books: a treatise on the branch of logic called "recursive functions" and a beautiful popularization of mathematics, *Playing with Infinity*. But in the Hungary of the 1930s, as a Jew, she was not eligible for a professorship at the university.

"Rózsa Péter was wonderful," he says. "She was immersed in mathematics. She was interested in how people think. The very first thing we did was to read *The Enjoyment of Mathematics* by Rademacher and Toeplitz. I was twelve or thirteen." This famous classic is perfect for a young beginner attracted to mathematics. It takes up a series of independently attractive topics in algebra, geometry, number theory, and logic in a challenging yet accessible style.

With Rózsa Péter he went through *The Enjoyment of Mathematics*, mastering numbers, curves, algorithms, and proofs galore. Sometimes she would ask him before they read a proof, "Can you do it yourself? Next week when you come, try to do it."

"I went to her house maybe twice a week for a couple of years, until I left Hungary at fifteen and a half. Sometimes she would take me to meetings of the Mathematical Society. I was the youngest. I was a little shy, but I went anyway. At one meeting someone was presenting Robbins' theorem on directing the edges of a graph so you can get from any point to any other point. The theorem says you can do that unless the graph can be disconnected by removing a single edge. The last lecture I heard was Hajós presenting his solution of the Minkowski problem about paving space. I didn't really understand it."

As the 1930s and 1940s advanced, two things in the life of the Lax family grew larger and larger. The first was young Peter's wonderful gift for mathematics. The second was horrifying anti-Semitism in both Hitler's Germany and Horthy's Hungary. By 1941 Hitler had conquered France, and war with Russia was coming next. What to do?

No place in Europe was safe. They would have to cross the ocean to America. And America wasn't very welcoming to refugees. The Great Depression wasn't over yet. There was resistance to foreigners who might take jobs away from Americans. And anti-Semitism was far from unknown in the U.S. It was considered advanced and liberal, that the patrician president, Franklin Delano Roosevelt, a distant cousin of former president Theodore Roosevelt, had chosen as his Secretary of the Treasury a Jewish banker, Henry Morgenthau. Some people called the president "Rosenfeld", as a term of abuse and hatred. Many European Jews trying to flee Hitler were unable to obtain U.S. visas.

Fortunately, Dr. Henrik Lax had patients and friends from the U.S. Peter says, "There was an American who had visited Hungary a few years before. He had been in an accident and my father had saved his life. He sent an affidavit. And also the American consul in Budapest was a patient of my father's."

It's not easy for a successful physician to abandon his patients and hospitals, and his prestigious, comfortable life to run to a strange country. Klara Lax insisted, demanded, that they escape while they could. Some who could have done so delayed too long. One such was Ferenc Polgár, Henrik's colleague at the Jewish Hospital, where he was chief of radiology. It was especially hard for a radiologist to abandon his cherished X-ray machines. Polgár and his family nevertheless did survive, fortunately for me.

"At that time," says Peter, "America to me was a child's dream. I really didn't know English, although I had studied it for two years. What did I know about America? Skyscrapers, tap dancing, Hollywood, chewing gum, and the electric chair. As children we were fascinated by the idea of an electric chair. I had read *The Last of the Mohicans, Tom Sawyer*, and *Huckleberry Finn*, and a novel by Jack London about two dogs that could sing—you know, sort of howl. I had also read *Helen's Babies* and *The Diary of a Bad Boy*. All were in Hungarian translations."

He says his English tutor in Budapest was wonderful. "He didn't assign childish reading; he assigned adult reading." Peter read the short stories of Somerset

Maugham, who was then widely read and highly respected. Maugham is not so much talked of anymore, but Peter says people should still read him; he is a first-class writer.

In 1940 Rózsa Péter suggested that Peter take the Eötvös Competition, even though Peter was only fourteen years old and it's restricted to high school graduates. His participation had to be unofficial. The regular participants sit in a closed room, watched by proctors, to work on three contest problems. The first question is usually easy, the last is never easy. Peter walked over to the university, found the contest room, received the three problems, took them home, did them, and turned them in the next day. All three answers were perfect.

Again the next year, 1941, at age fifteen, a few months before the Lax family departed from Budapest, he did it again. Unofficial participation and perfect score.

Here is the third problem from 1941: "ABCDEF is a hexagon inscribed in a circle. Suppose that the sides AB, CD, and EF have the same length as the radius. Prove that the midpoints of the other three sides form an equilateral triangle."

Sixty years later, during his acceptance speech for the Abel Prize in Oslo, Peter said that in his youth in Hungary, "Problem solving was regarded as a royal road to stimulate talented youngsters. I was very pleased to learn that here in Norway they have a successful high-school contest, where the winners were honored this morning. But after a while one shouldn't stick to problem solving. One should broaden out. I return to it every once in a while, though."

A few months after his second success on the Eötvös Competition, Peter joined a math study group at the Jewish Gymnasium. Laszlo Fuchs, a Hungarian-American algebraist now retired from Tulane University, was in that group with Peter. He writes, "At the beginning of the 1941–42 school year, Paul Erdős's father visited the principal of the Jewish Gymnasium in order to secure a classroom. The Jewish Cultural Association had organized a special course for high school students interested in mathematics, and needed a place to meet once a week. The instructor was Tibor Gallai, a very talented mathematician and a superb teacher (without a job). The school principal agreed, under one condition: his son (me) would also be admitted to the course. Peter Lax was the youngest one of 8 or 10 students in the group. Most of us were in high school. Some had already graduated, but because of the *numerus clausus* (quota against Jews) weren't allowed to enter the university.

"Before and after class we had long discussions on the solutions of the take-home problems, possible generalizations, and so on. Peter participated actively, and it was clear that he knew more mathematics than most of us. He left a couple of months later, when his family immigrated to the U.S. We were extremely sorry that he wouldn't come to the class any more, but we were happy for him that he could leave the country. He was one of the few in the course who would survive the Holocaust."

Indeed, the Lax family had tickets for the November 25 train to Lisbon.

Rózsa Péter and Dénes König wrote letters to mathematicians in the U.S., calling attention to Peter's exceptional promise.

Peter writes, "Dénes König wrote to John von Neumann, Gabor Szegö, and Otto Szasz asking them to look out for me. Only three years later he killed himself, to avoid arrest and deportation by the Nazis."

König's letter to von Neumann is dated November 12. Here is a translation into English:

My dear friend,

It is my understanding that Dr. Henrik Lax, a physician from Budapest, is moving with his family to America within the next few days. He is taking with him his son Peter Lax who is a gymnasium student in the VI grade. I have been in mathematical contact with young Peter for more than a year. Based on personal conversations and on his computational work (which consisted of solving basic mathematical problems) I have become convinced that the young man possesses extraordinary talent in mathematics. For instance, for two successive years, he succeeded in solving the tasks administered during the Mathematics and Physics Society's competition at a much higher level than those students who were the official participants. It should be in Peter's interest, but also in the interests of the community, if this exceptional talent were to be further nurtured and supported. Thus I am asking you, that if and when Peter Lax approaches you with some questions or requests for advice, would you please respond to him with the good will appropriate for a future scientist.

With great respect and friendship,
Dénes König
Budapest, XI
Horthy Miklós ut. 28

Three years later, in the days of Arrow Cross terror before the liberation of Budapest, "when the German army occupied Hungary, putting Hungarian Nazis in power, König saw what was coming and threw himself out the window of his apartment" (Lax, *Functional Analysis*, page 159).

So the Lax family got their papers and on November 25, 1941, they boarded the train from Budapest to Lisbon, Portugal, where they had tickets for a steamship to New York City.

As the Laxes traveled through Nazi Germany, they shared a train compartment with *Wehrmacht* soldiers. "When we reached the Swiss border," Peter later remembered, "the German guard checked our papers, and then said, 'Just a moment.' The air froze for us. But to our relief, he only asked if we still had the ration coupons for meat and butter that we had received on entering the Reich. My father gladly gave him the coupons."

Their ship departed Lisbon on December 5, 1941, the last U.S. passenger ship to leave Europe for the next four years. They were in the Atlantic Ocean when Japan bombed Pearl Harbor two days later. Now the U.S. was at war with the Hitler-Mussolini-Tojo axis, and Hungary, an ally of the Axis, was a wartime enemy of the U.S. But young Peter wasn't wasting his time on board ship. While they were crossing the Atlantic, he taught himself calculus. When they arrived in New York, they were at first classified as enemy aliens and taken to Ellis Island but released after a few days.

In the next four years, more than two thirds of Hungary's Jews, including many of the Lax family's relatives, friends, and neighbors, would be murdered by the Germans or by their Arrow Cross friends.

"We were the only members of my family who escaped the war in Europe. Of the ones who remained behind, my uncle Imre, my mother's brother, was killed while in a labor battalion. Another uncle, the husband of my mother's sister, Dr. Victor Farkashazi, and his son Steven, were murdered by Hungarian Nazis in Budapest. His wife and younger son escaped and eventually ended up in the U.S. His son is my cousin Bill Farr (he abbreviated the name Farkashazi), a retired tool-and-die maker. He worked for many years for General Motors. He has five talented children.

"When General Motors was contemplating reducing the pension and health care benefits of its retired workers, I asked my cousin if he was worried. He answered, 'I haven't worried since 1944.'

After the defeat of Hitler, Peter went every other year to visit Hungary, where he had three surviving cousins and mathematical contacts. His parents sent packages and money to their relatives and visited Hungary as soon as possible. Peter's father, Henrik, brought his mother to New York, where she lived for twenty years until the age of ninety-four.

"She was a remarkable woman," writes Peter, "an early suffragette. Her husband died when she was 38, and after that she supported herself and her children by a sewing business. She survived the war hidden in a hospital run by one of my father's closest friends, Dr. Géza Petenyi, whose name is inscribed in Jerusalem on the list of righteous gentiles.

"He saved the lives of others, one of them a very close friend of mine. He was being deported; but he managed to telephone Petenyi and tell him what road the deportees were on. Petenyi hailed a taxi and followed the column. When they made a rest stop, he got out of the cab, walked up to my friend, and said: 'Walk very slowly with me to that taxi over there and get in.' They got away."

But Peter's father, Henrik, never forgave the Hungarians for what they had done. When Henrik went back to Budapest to fetch his mother and visit his sister, he never stepped out of his hotel. After his sister died, he never went back again.

After the suppressed Hungarian revolution of 1956, the Soviets let Kadar liberalize the regime and allow people to travel to the West.

"My father's sister and her husband managed to get out in 1957 and settled in New York. They prospered; he was a very successful accountant. She lived to be 93 years old. There is longevity in my father's family; he lived to be 95."

CHAPTER 2

Manhattan, NY, and Manhattan Project. An army private among the "Martians"

In the U.S. the Lax family contacted Klara's cousin's husband, Gabor Szegö, at Stanford University. "My mother and Mrs. Szegö were first cousins, and our families had been friendly in Hungary. When we came to America, the friendship was renewed. They were very nice to me."

They also contacted von Neumann. "Von Neumann was always very kind to me." Von Neumann visited Peter at home and advised him to get in touch with his collaborator Francis Murray at Columbia University. But Gabor Szegö told the Laxes that Richard Courant at New York University was very good at working with young people. Peter later said, "That was the best possible advice."

Courant as a Jew had been expelled from his leadership of the famous mathematical school at Göttingen in Germany and had been working hard for several years to build a great mathematical center at New York University. He greeted Peter warmly and welcomed him into his family in suburban New Rochelle. The Courants had two sons, Hans and Ernst, and two daughters, Gertrude and Leonore (Lori for short). Peter was already then an enthusiastic tennis player and amusing conversationalist.

Although classified as enemy aliens, "Within a month, my brother and I were in high school." Peter signed up at Stuyvesant High, a very selective science-oriented public school. He joined the math team, of course. "We won the City Championship that year. Somehow mathematics is the same everywhere. There were other students I could talk to about mathematics. Two of them, Marshall Rosenbluth and Rolf Landauer, became very successful physicists and members of the National Academy of Sciences."

While still in high school Peter was invited by Paul Erdős to visit the Institute for Advanced Study in Princeton. Erdős introduced him to Albert Einstein as a promising young Hungarian mathematician. Einstein said, "Why Hungarian?"

He didn't take any math courses at Stuyvesant; he was far beyond their math teachers. "I had to take English and American history, and I quickly fell in love with America." The history textbook was illustrated by contemporary humorous cartoons about U.S. politics, far from the stuffy way national history was taught in Hungary.

His main job was to learn English, so he took two English classes. Both were classes in English literature. In one of them, he had to read Hardy's novel *Return of the Native*. To this day he doesn't know why or to what the native returned. As

Paul Erdős, 1991 (1913-1996)

Photograph by George Csicsery. All Rights Reserved

to the other literature class, he doesn't remember what they read. He passed his English courses with a C. He also had some more private tutoring in English.

Henrik and Klara enrolled Peter's older brother, John, in a well-recommended private high school. John got his high school diploma half a year later than his younger brother, Peter. John became a physicist. He lives in Washington, D.C., and has a literary bent, with a gift for translation. No doubt he bore a burden as the older brother of a famous prodigy. There is a certain parallel to the later strains and tensions between Peter's two sons. The older son, Johnny, was hyperactive, charismatic, and charming. Their younger son, Jimmy, struggled to defend himself against big brother's heavy teasing.

After graduating from Stuyvesant, Peter enrolled at New York University, where Richard Courant was making great progress in building his research and graduate program in mathematics. Undergraduate teaching was still mainly done by holdovers from the previous regime.

Dr. Lax opened an office in a fashionable neighborhood on the east side of Manhattan. For a few years the family lived on the East Side, and then in 1950 they settled into The Eldorado, a 31-story building on the West Side overlooking the lake in Central Park. (No one notices that the words "The" and "El" are redundant.) Their apartment was on the eleventh floor, facing the park. (It's now occupied by Peter's son, Dr. James Lax.) Two stories down, facing south, is the apartment where Peter and his wife, Anneli, would raise their sons.

The Eldorado is one of the Upper West Side's iconic architectural landmarks, filling the block front between West 90th and West 91st Streets. Its two majestic towers light up at night like giant candles. I have picked them out from the air as my plane approached LaGuardia Airport. The building is considered one of New York's finest art deco structures, with futuristic sculptural detailing and contrasting materials and textures. Among its tenants have been rabbi Stephen S. Wise, Barney Pressman (the owner of Barney's clothing store), Alec Baldwin, Faye Dunaway, Garrison Keillor, and Tuesday Weld. The novelist Sinclair Lewis lived in a tower apartment. In Herman Wouk's novel *Marjorie Morningstar*, The Eldorado is Marjorie Morningstar's address. A few blocks north are the Central Park tennis courts, twenty-five blocks away is Lincoln Center with the Metropolitan Opera, and NYU is a subway ride downtown. The family has had season tickets at the Met for many years.

The dining room in Peter's apartment has comfortable chairs at the table, about six side chairs against the walls, several large green plants in big pots, paintings on the walls, framed photographs on side tables, and a small balcony, where you look down at the interior courtyards of the block west of Central Park. In the living room there is plenty of space for the baby grand piano, and for bookshelves displaying works about art, literature, history, politics, and philosophy. There are large paintings on the walls and a small balcony looking out over West 90th Street. Of course there are also bedrooms, a kitchen, a pantry, and a maid's room.

Into the army

Peter Lax escaped from the European firestorm. However, the U.S. entered the war just as he arrived, and in 1944, after less than two years of college, he became eighteen, draft age. But he did not become cannon fodder in the U.S. wars against the Germans and the Japanese. After the unavoidable basic training (which he survived well, being in good physical shape) Peter was sent to school at Texas A&M. From there, after an interlude at Oak Ridge, he was shipped in June 1945 to the best possible place for him—a secret physics laboratory, where he would work alongside the world's greatest scientists, including John von Neumann.

Just how this came about Peter never knew for sure. He suspected the machinations of Richard Courant. "Courant had written Oswald Veblen to see what could be done about me. Because of his own experience in the trenches in the First World War, he was very concerned about the possible loss of talented young people."

Richard Courant was the man who provided the Mathematics Institute at Göttingen with a handsome new building, thanks to the Rockefeller Foundation. Later on it would be he who would make NYU's graduate math program the main mathematical support for the U.S. military. And a few years later, it would be he who would win a grant from the Sloan Foundation (aka General Motors) to build a 13-story headquarters for the Courant Institute. Who then but Richard Courant would have the know-how to induce the U.S. Army to pluck out a certain Private Peter D. Lax from among millions of trainees and assign said Private Lax to a certain unknown hilltop northwest of Santa Fe? Indeed, in the NYU archives I found copies of letters from Courant to influential mathematicians, trying to get Peter a good assignment while in the U.S. Army.

Peter writes, "I was first sent to a camp in Florida—an Infantry Replacement Training Center. A very ominous sounding description, the army didn't mince words. I was thin but very strong, I could handle a rifle, and I had no trouble on the marches. I got along with the people in my barracks. They were a cross-section, including some hillbillies, who I am sure found me as strange as I found them.

"I missed 'Village Fighting' (a regular part of basic training) because I was sent to ASTP, the Army Specialized Training Program. It was a program to take out of the ranks people who had some college education or had scored high on the army intelligence test. I was sent to Texas A&M (agricultural and mechanical) University in College Station, Texas, to learn some applied mathematics that might be useful to the military. Today it is a very good school, but then it was the butt of a lot of 'aggie' jokes. All male, no blacks, a very antiquated faculty, with some exceptions. Everybody in uniform. I passed the calculus test with flying colors, so I was excused from taking that course. One of the few research mathematicians on the faculty, Dr. Edmund Chester Klipple, offered to run an R. L. Moore-type seminar for me and another GI, Sol Weinstein, who also passed his calculus test with flying colors."

At that time every research mathematician at any southern university was a Moore student. Klipple's 1932 dissertation at the University of Texas was "Spaces in which there exist contiguous points".

"We proved theorems in real variables, some of them quite sophisticated. I learned quite a lot, for instance Euler's formula for the buckling of a column. This idyllic life came to an abrupt end when I received orders without further explanation to join 'the Manhattan project'; all I knew about it was that it wasn't in Manhattan." Peter had no idea that his family friend Leo Szilard, through a letter to Roosevelt signed jointly with Albert Einstein, had initiated a project to build an atomic bomb in the USA.

Manhattan Project. An army corporal among the Martians

"I got orders to go to Oak Ridge. Then one morning in Oak Ridge, about four weeks later, we are ordered to fall out and pile onto a train for an unknown destination."

("Fall out" is the army's command for "break out of your military formation, go individually to your next assignment.")

"After a couple of days we arrive at a place called Lamy, New Mexico," an insignificant location which, for obscure historical reasons, is the closest the Santa Fe Railroad gets to the city of Santa Fe.

(L–R): John von Neumann, Richard Feynman and Stanislaw Ulam on picnic in Bandelier National Monument, New Mexico, 1949

"In Lamy we fall out again and pile into buses. We are driven to a small city. I recognized Santa Fe, for in 1942, our first summer in America, my parents, my brother, and I drove across America and passed through Santa Fe. From Santa Fe we are driven into the Jemez Mountains, into nowhere.

"It is then that I realize we are going to a secret project. We reach the secret site called Los Alamos."

There Peter lived in a barracks, like any regular army soldier, along with other enlisted men who had been selected for Los Alamos. Some had done undergraduate work in physics or math, others were skilled machinists. While in the barracks they were subject to ordinary army discipline, including Saturday morning inspections.

"A few days after our arrival at Los Alamos we are told the task of the laboratory: to build a nuclear bomb whose explosive energy 'e' comes from Einstein's $e = mc^2$. The bomb is built from plutonium, an element that does not exist in the universe, except for what we manufacture at a nuclear reactor at Hanford, Washington. We are also told that the energy from a nuclear reaction can be used not only for bombs but also to supply all the energy the world needs at very low cost. You can imagine the impact of these revelations. Partly because of that experience, I have never cared for science fiction. I had lived it.

"It was like a dream. The army didn't have much power over us; it was the first time I was really on my own. There was the pleasure of being in a group working on a specific project. There was one goal, and I realized where mathematics fit into it.

"The secrecy of Los Alamos was secured partly through rigorous security rules, but also because the idea of atomic weapons was so unbelievable. However, one of my friends, the eccentric mathematician Paul Erdős, who had a very broad circle of friends and acquaintances, guessed what was going on. He wrote me a postcard:

'A little birdie told me that you are working on atomic weapons.'

But since he wrote in Hungarian, he did not use the English idiom 'a little birdie told me', but rather its Hungarian equivalent: 'My spies report to me'. Fortunately he sent the card to my home address, and my parents wisely did not forward it.

"Von Neumann was around a lot. He was a very important person already then, and extremely busy, but whenever he came he would give a lecture. At that time he talked about game theory and computers. He was very much convinced that the task at Los Alamos would turn toward computing, since they would be unable to fulfill their mission without it.

"No mere list of von Neumann's achievements gives a proper picture of the man; for those who are too young to have glimpsed him I offer the image of Gelfand and Michael Atiyah rolled into one, with a couple of physicists and economists added for good measure. He carried thinking further than most people can conceive of its being carried. That is the reason he was so much sought after by the government for advice. Not only the government but many mathematicians sought von Neumann's advice about their research. Not that von Neumann was able to solve a difficult problem in a single interview, but he had an uncanny ability of relating it to other problems. Often such a reformulation represented the labor of six months of the person who posed the question.

"Richard Feynman was also there. Although he was quite young, he was already legendary. I had just turned nineteen, so my interaction was more with young mathematicians like Richard Bellman, John Kemeny, Alex Heller, and Paul Olum. We became friends, friendships that lasted a lifetime. I also saw a lot of Stan Ulam."

These young mathematicians, and Peter himself, had been picked by an army program specially designed to detect and utilize such "brainiacs". "An important supplement to the prestigious scientists who were recruited were the draftees of the MED's (Manhattan Engineering District's) Special Engineer Detachment (SED). The SED was composed of mostly young scientists, engineers, and technicians who had been drafted into the army. Rather than being sent off to combat, they were transferred to Los Alamos, or elsewhere, to fill a personnel shortage and perform a variety of scientific and technical tasks. Almost 30 percent of them had college degrees. Many had been in graduate school when they were called, and some had completed their Ph.D's. By the end of 1943 nearly 475 SED's had arrived; by 1945 the unit included 1,823 men." (*Racing for the Bomb*, Robert S. Norris, Steerforth Press, South Royalton, Vermont, 2002, page 247.)

Peter writes that "some of the leading scientists, Oppenheimer, Bethe, Kistiakowski, Teller, and others were the leaders of the Project. World famous scientists like Fermi, von Neumann and others were frequent consultants. I wondered how come that I was chosen for this project. I surmised, and still believe, that Courant had recommended me."

John von Neumann (left) and Manhattan Project leader Robert Oppenheimer (right) in front of Princeton's Institute for Advanced Study computer, 1952

Niels Bohr (left) and Richard Courant (right)

Hiking and cross-country skiing were great weekend pastimes. He has a photo (shown here) of himself seated on a mountaintop near Enrico Fermi, the leading physicist of the Manhattan Project. He even played tennis with Fermi. He thinks he won. The memory still makes him smile. He tried to learn to fly an airplane. At age eighteen, while still an army private, he published his first paper, the proof of a conjecture he had heard from Paul Erdős, "On the derivative of a polynomial," in the *Bulletin of the American Mathematical Society*.

"I was placed in T (for Theory) Division. There was no special mathematics group, perhaps because the leading mathematician at Los Alamos, Stan Ulam, was not inclined to do administrative work. So mathematicians were assigned to work with physicists; that way we learned how a mathematician can help solve a scientific problem.

"At first I was asked to do calculations with a Marchant, a hand-operated electric calculating machine, but very soon I graduated from that. I knew very little physics then. I still wish I knew more. I was assigned to work on neutron distribution in general, neutron transport problems. The fanciest piece of work I did was a calculation of the criticality of an ellipsoid. The theory of neutron transport is a linear theory. Many years later I heard the distinguished French physicist Robert Dautray, who was one of the leaders of the French nuclear development, remark that since nuclear fission is governed by linear laws, it is possible to scale up experiments. Nuclear fusion, in contrast, is governed by a nonlinear theory, so it is not possible to scale up experiments. This is one of the difficulties that makes it so hard to generate energy by nuclear fusion."

This assignment at Los Alamos—to calculate the critical size of an ellipsoid, as a generalization of a sphere—was a rather classical bit of applied mathematics. The ellipsoid in question would be composed of "fissionable material"—uranium 235 or plutonium. "Critical size" means the size needed in order to sustain a chain reaction. A sufficiently eccentric ellipsoid might serve to represent a cylinder. One of the bombs that would be dropped on Japan was cylindrical in shape.

The prominence of four brilliant Hungarians in the Manhattan Project (Eugene Wigner and Leo Szilard at Argonne Laboratory in Illinois, and John von Neumann and Edward Teller in Los Alamos) resulted in the following story: "Oppenheimer managed to secretly recruit four extraterrestrials for his lab, but they could be recognized by their weird English accents. As a coverup, they all claimed to be Hungarian."

To tell the truth, Peter Lax has never completely lost the last vestige of his Hungarian accent.

At the top of the list, in terms of sheer brain power, was John von Neumann. "He had interviewed me in New York soon after we arrived. After we met again at Los Alamos, he was always available, even though he was already a very important person and extremely busy. When I latched on to some of his contributions, like the von Neumann criterion for stability and shock capturing, he was interested in my ideas, and I profited from his ideas." (More on this in Chapter 10.)

Peter has written a short biography of von Neumann, which I include as Appendix 2, both to present the story of his life and accomplishments and to convey the tremendous admiration for von Neumann that Peter has felt all his life. The last

Gabor Szegő and George Pólya, Berlin, 1925

Courtesy of Gerald L. Alexanderson, from Pólya Picture Album: Encounters in Mathematics

few paragraphs describe von Neumann's foresight about computing and his seminal work on shock waves and the approximate solution of partial differential equations. Peter could have added, "and much of my own work has been a continuation and development of von Neumann's beginnings in these areas."

Among mathematicians, Peter has a great reputation as a storyteller. He has many stories about the young mathematicians he met as a nineteen-year-old army private at Los Alamos.

"Dick Bellman was somewhat older than the rest of us; he was already married. His wife moved to an apartment in Albuquerque to be near Dick. The expense proved too great, so she joined the WACs (Woman's Army Corps). Dick sublet the apartment to a fellow soldier named Greenglass, who a few years later in 1950 became infamous as an atomic spy, brother-in-law of the Rosenbergs. The transmission of atomic secrets took place in the apartment that Greenglass had leased from Bellman. By this time Bellman was consultant to the Rand Corporation on a classified project requiring secret clearance. The connection with Greenglass caused

Peter (left) and Enrico Fermi (right) on a hike in the mountains around Los Alamos with a colleague

the suspension of his clearance pending a security hearing. I was able to testify that far from being a friend, Dick loathed Greenglass. His clearance was restored.

"Many years later Bellman was felled by a brain tumor; it was not malignant but caused extensive damage, including to muscles of his face. Dick said that it made him look like a Picasso portrait. He didn't think he had lost any intellectual brain function, 'but if I did,' he said, 'I could afford it.'

"John Kemeny was another of the prodigies who made it to Los Alamos. He was a brilliant student at Princeton when he was drafted, and captain of the debating team in his freshman year. He came to the U.S. with his family from Hungary a couple of years before I came, and after being drafted received U.S, citizenship, as did I. He wrote his parents how pleased he was to have this honor bestowed on him so soon and without being asked any stupid question about being a communist. All mail in and out of Los Alamos was censored, so the remark about being a communist was picked up, and John was summoned by the base Security Officer. His friends were waiting anxiously in the lobby of the security office while the interrogation took place. After about two and a half hours John emerges. 'What happened?' we asked. 'I made the Security Officer admit that there are several things wrong with the capitalistic system,' John said."

Kemeny has provided a vivid description of computing as it was done in the Manhattan Project, with von Neumann's response. (*Man and the Computer*, quoted on page 275 of Marx.)

"In Los Alamos each of our machines was designed to carry out one or two arithmetical operations, for example $A \times B + C$. The values of A, B and C were fed to the computer by a deck of IBM cards, and the computer performed the same operation on each card. The new deck was then moved manually to the next calculating machine, which carried the calculation one step further. The control of the machine was by means of a plug-board, which had to be specially rewired for each type of operation. Neumann argued that those machines, which depended heavily on mechanical parts, were much too slow to be useful. Therefore he proposed an entirely electronic device. He went on to argue that while the decimal system was perfectly practical for mechanical devices, a binary system would be much easier to implement electronically because of the efficiency of an on-off system. Next he pointed out that if we had faster machines it did not make sense for human beings to have to interfere at each step. Therefore he advocated the existence of an internal memory in which partial results could be retained so that the computer could automatically go through many rounds of operations."

Kemeny became chairman at Dartmouth, where he developed the computer language BASIC and computerized the campus. He wrote a seminal introductory text, *Finite Mathematics*, with Snell and Thompson. He was chosen to be president of Dartmouth. During a ten-year tenure he accomplished a lot, against much opposition by alumni: the admission of women and many reforms of the curriculum. He told Peter that when the trustees offered him the presidency, they warned him that "this is a 100% job, that he should forget about his former interests." John told them that he realized that, but relaxing is important too, such as playing golf. Would the trustees mind if twice a week he took off a couple of hours. Not at all, they told him. "Actually I don't play golf," said John, "but I would like to teach a math class."

Stan Ulam, one of the originators of the so-called Monte Carlo method of calculation, was one of von Neumann's closest friends; that is how he came to be at Los Alamos. Peter says that Ulam was a most unusual mathematician: "He worked with ideas, not with equations or formulas. He had an original sense of humor; when after the war Los Alamos was somewhat opened up to the outside world, the Catholic community installed a prefabricated church in town. Stan dubbed it 'Santa Maria della Bomba'."

Even Niels Bohr visited once; since he was so closely associated with nuclear physics, the security people insisted that he use an alias, Nicolas Baker. (Enrico Fermi's alias was Henry Farmer.) At a party in the evening one of the guests who had been to Copenhagen in the years before the war recognized Bohr and said, "Professor Bohr, how nice to see you here."

Bohr, remembering the drilling by the security people, replied, "No, I am Nicolas Baker." But then he added, "But you are Mrs. Hautermans."

"No", she replied, "I am Mrs. Placzek." (She had divorced and remarried.)

When asked about the Hungarian physicist Edward Teller, Peter replied, "Well, of course blood is thicker than water, even heavy water, so we were friends." (They were both graduates of the same gymnasium, the *Minta*.) Teller later became "the father of the H-bomb" and also the principal advocate of the "Star Wars" project of antiballistic missile defense against a possible nuclear attack, which was embraced by President Ronald Reagan despite the nearly unanimous judgment among competent experts that it was a dangerous delusion. Teller also led in labeling Robert Oppenheimer, the main leader of the Manhattan Project, as a "security risk."

But all that happened later on. To Peter as an army corporal far from home, Teller was a welcome familiar face.

At the time, nearly everyone involved with the atomic arms race was deeply worried that the Germans would beat us to it. They had Werner Heisenberg and a whole crew of other world-class physicists. In fact, during the war Heisenberg came from Germany to visit Bohr in Denmark, hoping to learn in conversation with Bohr whether the U.S. was working on an atom bomb. He didn't learn any such thing. But as a result of this visit, Bohr did realize that Germany was indeed working on it.

The physicist Frank Hoyt, one of Peter's supervisors at Los Alamos, was charged with scrutinizing all publications of Heisenberg to see if they were a by-product of work on atomic weapons. Hoyt concluded that Heisenberg's basic paper on scattering theory, published in 1943, had nothing to do with nuclear weapons. On May 8, 1945, Germany surrendered. It became clear that despite their head start they had never come close to making an atom bomb. Werner Heisenberg claimed that he had slowed down the project. Niels Bohr doubted this.

Two types of atom bomb were being built: one with uranium 235 and one with plutonium. The uranium bomb was relatively simple; there was no necessity for a test. The plutonium bomb was more complex, so one was set aside for testing. Six weeks after Peter arrived at Los Alamos, on July 16, 1945, the plutonium bomb was tested at a remote desert site called "Trinity" in the Alamogordo Bombing and Gunnery Range.

A little after midnight, the physicist Donald Hornig climbed down from the tower where the bomb was placed and went to a bunker five miles away, where he joined Robert Oppenheimer and others who were waiting to see if the "the gadget", as they called it, would actually go off. Hornig took his next assigned position: keeping his finger on a control switch to abort the blast if something should go awry. Unexpected thunderstorms delayed the experiment. But the weather cleared up, and at 5:29 a.m. the bomb exploded. Hornig later said that the swirling orange fireball filling the sky was "one of the most aesthetically beautiful things I have ever seen." (See the photo following this chapter.) The explosion was much stronger than the physicists expected; some of their measuring instruments were destroyed by the blast. In the following days, Peter witnessed "the frantic effort of scientists to calculate the effect of the explosion, and their efforts after the test to reconcile the measured values with their calculations."

With Nazi Germany defeated, the original impetus that had led Leo Szilard and Albert Einstein to urge the bomb on Franklin D. Roosevelt no longer existed. The war against Japan was continuing. Many of the physicists who had worked on the bomb were opposed to dropping it on Japanese civilians. Szilard circulated a petition urging a demonstration of the bomb on some uninhabited place to give Japan a chance to surrender without suffering an atomic explosion. Lewis Strauss, an influential banker and admiral, likewise opposed dropping the bomb on a Japanese city without first providing a demonstration of its power.

Peter Lax, like other enlisted men in danger of being sent to invade Japan, was in favor of using the bomb. "I was in the army, and all of us in the army expected to be sent to the Pacific to participate in the invasion of Japan. You remember the tremendous slaughter that the invasion of Normandy brought about. That would have been nothing compared to the invasion of the Japanese mainland. You remember the tremendous slaughter on Okinawa and Iwo Jima. The Japanese would have resisted to the last man. The atomic bomb put an end to all this and made an invasion unnecessary. I don't believe revisionist historians who say: 'Oh, Japan was already beaten, they would have surrendered anyway.' I don't see any evidence for that."

He raises another point. "Would the world have had the horror of nuclear war if it had not seen what one bomb could do? The world was inoculated against using nuclear weaponry by its use. I am not saying that alone justifies it, and it certainly was not the justification for its use. But I think that is a historical fact."

The generals running the Manhattan Project and President Harry Truman had little respect for people with qualms about the bomb. In Truman's inner circle, failure to use every available weapon against Japan would have been considered politically fatal. Calculation about a presidential election would have trumped any other consideration.

Peter writes that "members of the team that was sent to the Pacific to help arm the bomb received an inoculation against yellow fever, which causes a somewhat painful swelling of the arm. Since their mission was top secret, they were sternly warned not to cry out if someone accidentally bumped into their arm.

"When the uranium bomb was delivered to the air force base in the Pacific to be dropped on Hiroshima, scientists from Los Alamos explained to the commanding officer, General LeMay, the basic principles of nuclear fission. 'How many times have you tested this bomb?' LeMay demanded to know. The bomb measured only 28 inches by 120 inches. When told that it had never been tested, he laughed out loud and regarded his visitors as irresponsible eccentrics."

At 8:15 a.m. August 6, 1945, at an altitude over Hiroshima chosen to be optimal for maximal effectiveness, a calculation credited to John von Neumann in his U.S. government award for military service, "there was a blinding flash in the sky, and a great rush of air, and a rumble of noise extending for many miles" (Official U.S. government report of the Manhattan Project Atomic Bomb Investigating Group).

Around 50,000 died immediately, mostly "noncombatants"—women, children, and men too old for military service. As many as 166,000 more perished within four months because of the bomb. Others died later from cancers induced by nuclear radiation.

The emperor of Japan consulted with his generals about this new American weapon. The Soviet Union was joining the war against Japan, massive numbers of Russian troops were moving from the European front to the Pacific. To hasten the Japanese surrender, the U.S. dropped another atom bomb on August 9, this one over Nagasaki; 60,000 to 80,000 more "noncombatants" were killed instantly. Japan surrendered on August 15.

The explosions at Trinity, Hiroshima, and Nagasaki all took place before Peter had completed his analysis of an ellipsoidal bomb. As usually happens to such research, his report was first "classified" and then appropriately filed away.

Leo Szilard was the most active of many physicists who were deeply frightened of an atomic arms race. The U.S. had a monopoly on the bomb at first and tried hard to keep its manufacture a "secret", but scientists knew that in a few years Russia would have the bomb too. It turned out that the German refugee physicist Klaus Fuchs and others had already made Stalin aware of the Manhattan Project. The only secret was whether the bomb would work, and Hiroshima ended that secrecy.

A serious effort was made to achieve international control of atomic weapons through a United Nations agency. The Baruch Plan would have required both the U.S. and the U.S.S.R. to open up their secret atomic research to an international agency. The Soviets refused first.

The next big struggle was to take the atom bomb away from the U.S. Army and put it under civilian control. This effort was successful; the Atomic Energy Commission was created. John von Neumann became a commissioner.

Most of the leading physicists at Los Alamos quickly dispersed back to regular campus life. But a history-making laboratory such as Los Alamos could hardly be disbanded or destroyed. There remained plenty of interesting scientific research about this new physical phenomenon, the atom bomb. (Not to mention practical military questions: making more efficient, more convenient, more versatile bombs.) Los Alamos became a permanent site for science and engineering related to atom bombs, and two more such laboratories were created in Albuquerque, New Mexico, and Livermore, California. At Livermore, near Berkeley, Edward Teller had free rein to try to build his "Super", which had been opposed at the Manhattan Project by Robert Oppenheimer. Ironically, the ultimate success of the H-bomb was made possible by an insight contributed by the Los Alamos mathematician Stan Ulam.

There was a succession of atomic explosions, as various nations announced their entrance into the atomic club. On March 1, 1954, at Bikini, the U.S. performed a successful test of a hydrogen bomb. There is now an international agreement not to test atom bombs in the atmosphere, but underground testing is still permitted.

For several years after the war, Peter Lax and John von Neumann were visiting consultants at Los Alamos. Sandia National Laboratory has long been the major employer in Albuquerque, New Mexico, where I taught mathematics at the University of New Mexico. Between Sandia and Los Alamos, the atomic industry is a major source of federal dollars for New Mexico. Our senators and congress-persons, in bipartisan fashion, reliably and earnestly support its appropriations.

Szilard worked persistently to prevent a nuclear holocaust. In 1962 he helped to found the Council for a Livable World. The council's goal is to warn the public and congress of the threat of nuclear war and encourage rational arms control and nuclear disarmament. As the first physicist to envision a chain reaction and as the initiator of the U.S. effort to build an atom bomb, Szilard may have felt a special sense of personal responsibility for avoiding a nuclear catastrophe. While Peter was in Los Alamos, Szilard was working with Fermi at the Chicago branch of the Manhattan Project.

Asked by an interviewer who was his favorite among the so-called "Martians", the great Hungarian scientists who worked in the Manhattan Project, Peter answered, "Szilard had perhaps the most fantastic imagination. He could see the future and act on it. Very few people foresee the future and those who do don't do anything. But he did. Perhaps he was the most remarkable among them." The following story illustrates Szilard's foresight: The day after Hitler came to power, he withdrew his money from the German bank and left the country.

The physicists at Los Alamos, who had been kept from academic work, organized a Los Alamos branch of the University of New Mexico. Peter took some courses and was assistant in a course on mechanics given by Chaim Richman. "I wrote up the notes. I didn't know too much, but I was learning. The army still needed us at Los Alamos, so they made a deal. They agreed to discharge us if we agreed to stay on as civilians until the following May. It was a very good deal." Peter was discharged from the army with the rank of corporal.

In the summer of 1946 he went from New Mexico to California and spent another summer semester at Stanford with the Szegös. He writes, "This time I took some reading courses with George Pólya. Pólya's older brother was a good friend of my father's. He was a very prominent surgeon, very skillful and very innovative. In fact, George Pólya was known in Budapest as the talented kid brother of the famous surgeon Jenö Sándor Pólya." (Jenö Pólya is still remembered for the "Reichel-Pólya operation", a type of posterior gastroenterostomy. Between World War I and World War II, American surgeons came to Budapest to observe his surgical technique, and in 1939 he was elected an honorary member of the American College of Surgeons. Jenö Pólya was murdered by the Nazis during the Siege of Budapest. His body was never found.)

In Peter's article "The bomb, Sputnik, computers, and European mathematicians", he wrote, "It was under the leadership of Szegö that Stanford established itself as one of the foremost schools of analysis and statistics."

After getting his Ph.D., Peter returned to Los Alamos in 1949 as a staff member for a year. He writes, "By that time von Neumann had started to spend a lot of time there. The focus of his interest was numerical schemes. We didn't have a computer yet, but already during the war von Neumann realized that to do bomb calculations, analytical methods were useless; one needed massive computing. He also realized that computing is good not just for bomb-making but for solving any large-scale scientific or engineering problem. And he also realized that it's useful above all to explore which way science should be developed. As he said, 'It gives us those hints without which any progress is unattainable, what the phenomena are that we are looking for.'"

Peter coined a capsule summary of that insight:
Computing is the tool for the theorist, used in the manner of the experimentalist.

The fundamental mathematical-physical problem for theorists at Los Alamos and in laboratories around the world is a classical one inherited from the nineteenth-century: compressible flow. When studied from the viewpoint of heavier-than-air flight, it is known as "aerodynamics". When studied with interest in the earth's atmosphere on a large scale, it is known as "meteorology". When it is complicated by strong effects of electromagnetism, as in the interior of stars, it becomes "astrophysics". In attempts to create energy by nuclear fusion or in studies of the details of a nuclear or thermonuclear explosion, it is called "magnetohydrodynamics" or "plasma physics". It is "classical" or "old-fashioned" in the sense that the physics is primarily continuum mechanics, not quantum mechanics or relativity. The mathematics doesn't require "modern" innovations such as abstract algebra or algebraic topology or category theory. The governing equations would be understandable to Euler or Maxwell. But it still frustrates and puzzles the most powerful living mathematical minds. While the shocks—discontinuities in finite time—are a prominent phenomenon of *compressible* flow (aerodynamics), it is unknown to this day whether blowup in finite time is possible in the better-behaved case of *incompressible* flow (such as water waves). The Clay Foundation is offering $1,000,000 for a rigorous solution to that question.

These equations absolutely require high-speed computer simulation to go where pure analysis cannot penetrate. How to keep track of the shocks that can arise from a smooth initial state? This has been a continuing theme of Peter Lax's research. We discuss this work in more detail in Chapter 10. "When I got back to Los Alamos in 1949 there was a great deal of research on difference schemes for partial differential equations. But it wasn't until 1952 that von Neumann's machine, the JOHNIAC or MANIAC, was built. Then versions of it were built simultaneously in many places: at the Institute for Advanced Study in Princeton, at Argonne Lab in Illinois, and at Los Alamos, where Nicholas C. Metropolis and James Richardson were most deeply involved. That was when I started my experiments on shock capturing. I remember writing machine code for the MANIAC." "Experiments" here means "computer experiments", "numerical experiments". To study shocks "experimentally", the mathematician starts out with the differential equations for compressible flow. Since the computer works only with discrete information, he "discretizes" the system of differential equations, replacing the differentials or derivatives by finite differences. This first step is far from routine! It takes mathematical finesse to make an approximation which is amenable to computation and also reasonably true to the continuous flow it is supposed to approximate. Shocks can develop spontaneously, even if the original state was smooth. And it isn't easy to know where and when the shocks may arise! Then it is complicated to follow them as they move and evolve. "Shock capturing" was the method proposed by von Neumann. One uses some mathematical device to smooth out the flow, replacing discontinuities with regions of rapid transition, which are much easier to work with. By numerical experiments, von Neumann and Lax familiarized themselves with shock flows. This prepared them to devise effective computational algorithms and even rigorous proofs. Peter came back to Los Alamos every summer after that. He recalls, "Each summer there was a brand new machine which made last year's model primitive. Of course all these machines were incredibly primitive by our standards, but they could do more and more and more." During an interview, the mathematician Phil Colella seemed amazed that Peter Lax had actually written computer programs in "machine language" in the days before compilers and executive programs existed. *Colella*: So you were experimenting with shock capturing at Los Alamos in the late 1940s—well, I guess when the first computer arrived.

Lax: Yes.

Colella: You actually got in there and programmed these naked machines.

Lax: I actually programmed, yes. Later on I used assembly language, but still pre-FORTRAN.

Colella: What was the environment like there just after the war?

Lax: I would say ideal. The military realized that they had blundered by doubting the importance of science, and they decided never again to expose themselves to such danger. Their respect for science grew enormously, enormous resources were put at the disposal of scientists, and they were wise enough to see that they didn't always have to be devoted to a specific task; general research also had to be supported. That spirit still lives on, although not nearly to the same extent. "The time I spent in Los Alamos, especially the later exposure, shaped my mathematical thinking. First of all, it was the experience of being part of a scientific team—not just of mathematicians, but people with different outlooks—with the aim being not a theorem but a product. One cannot learn that from books; one must be a participant. For that reason I urged my students to spend at least a summer as a visitor at Los Alamos. Secondly, it was there, in the 1950s, that I became imbued with the utter importance of computing for science and mathematics. Under the influence of von Neumann, for a while in the 1950s and the early 1960s, Los Alamos was the undisputed leader in computational science."

NYU and the Courant Institute

In the fall of 1946 Peter went back to NYU, first as an undergraduate, then as a graduate student. The math program had just moved into the Bible House on Astor Place near Cooper Union. Back in 1853 the American Bible Society had relocated there from a modest headquarters on Nassau Street. The new grand and fashionable 5-story cast-iron structure was an architectural and technological marvel, occupying a full city block. It was financed by the richest Christians in New York, and it made the American Bible Society one of the most powerful reform organizations in the nation. Thousands of Christian tourists visited every year, and Mark Twain said after a visit to the Bible House, "I enjoyed the time more than I could possibly have done in any circus." The Bible House was primarily a publishing facility, and it produced tens of millions of bibles in many languages. It helped bring publishers, libraries, and bookstores to the Cooper Square-4th Avenue neighborhood. Its demolition in 1956 marked the beginning of the end of "book row".

From the *New York Times* (April 2, 1956)
1853 BIBLE HOUSE TO BE DEMOLISHED
Wreckers Start Work Today on City's First Building with Iron Framework

As a graduate student at NYU Peter found outstanding teachers, including Richard Courant, Kurt Friedrichs, and Fritz John, and outstanding fellow students, including Joe Keller, Cathleen Morawetz, Harold Grad, Louis Nirenberg, Martin Kruskal, and Anneli Cahn, soon to become Anneli Lax. Keller became a professor at NYU and then at Stanford, and the leader of a very prolific school of applied mathematics, especially known for his method of calculating diffraction of light rays from edges, corners, and cusps of reflecting obstacles. Grad became the key leader at NYU of a long-continuing group effort to understand hydromagnetic flows, or "plasmas". Controlling these is the key to generating electric power from nuclear fusion, the source of the sun's power. Kruskal became famous for an original way to represent relativistic mechanics and then later on for his role in relating the Korteweg-de Vries partial differential equation (the "soliton" equation) to the Fermi-Pasta-Ulam phenomenon (recurrence in a flow which was expected to be ergodic). Nirenberg, from Montreal, became Peter's lifelong closest friend and also, like Peter, a preeminent world leader in his own particular branch of mathematics, elliptic partial differential equations, which can describe equilibrium situations in elasticity, fluid flow, or electromagnetism.

Peter received his Ph.D. in 1949 for a thesis, "Nonlinear system of hyperbolic partial differential equations in two independent variables", written under Courant's student Kurt Otto Friedrichs. Partial differential equations (PDE) would be Peter's lifelong mathematical milieu. According to the Mathematics Genealogy Project, Peter by now has had 55 doctoral students and 518 doctoral descendents.

Leo Szilard (1898-1964)

John Kemeny, 1984 (1926-1992)

Stanislaw Ulam (1909-1984) with "Fermiac" created by Enrico Fermi

The Trinity Test, photographed just outside Base Camp, July 15, 1945. The photograph provided the basis for the Theoretical Division's earliest calculations of the Trinity weapon's yield.

Family Album

Henry Lax with sons Peter and John on an outing in the Grunewald, Hungary, 1931. Maternal Grandmother Ilka Kornfeld, née Nemenyi, watches in the background.

Henry and Klara Lax

The Eldorado–seen from across the Reservoir in Central Park–has been home to Peter's parents, Peter and his son James, since 1954. Jacqueline Kennedy Onassis, for whom the Reservoir is named, was a patient of Peter's father Henry.

Grandpa Henry with Jimmy on his knee, 1957

Peter and his father, Henry

Grandma Klara reading to Jimmy, 1957

Peter and his brother, John, at Prince Camp, Loon Lake

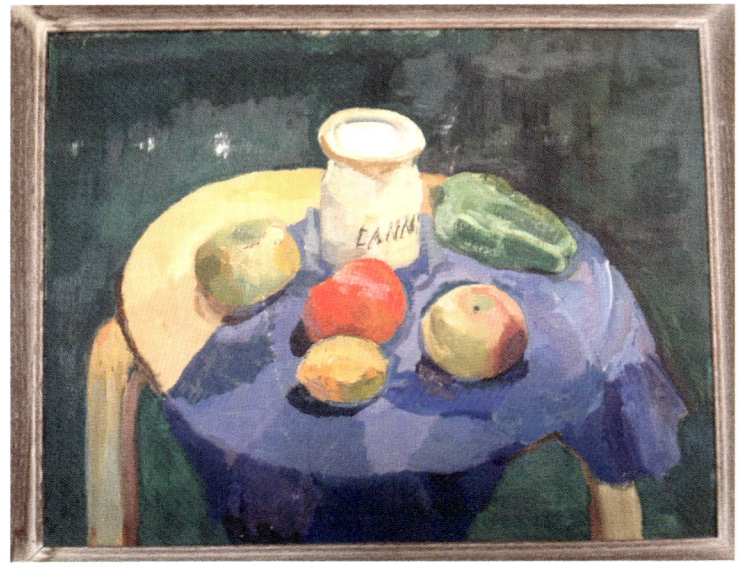

Still life painting by Peter Lax

Peter with his young son Johnny

Courtesy of Peter Lax

Peter with his young son Jimmy

Courtesy of Peter Lax

Peter giving Johnny tennis advice

Peter and Jimmy

Anneli and Peter at Courant Institute

John and Marnie, Greece, 1978

Photograph by James Lax

Peter playing tennis

Courtesy of Don Albers

Elizabeth Thomas, the Lax family nanny, with Jim Lax on his graduation from Columbia University, 1976

Courtesy of James Lax

Peter on the phone with the Norwegian Academy of Science and Letters learning of his award of the 2005 Abel Prize, 6:30 a.m.

Lori and Peter with Pierre at his college graduation, 2013

Photograph by James Lax

Peter with family and guests at Prince Camp, Loon Lake

Family Thanksgiving, 2013. L–R: Sabah D'Agrhi, Tommy, Timmy, Jim, Pierre, Peter, Nancy Foldi, Lori

CHAPTER 3

Family life: Son, husband, father, grandfather

I never met Peter's parents. "My mother died in 1973, at age 78, and my father in 1990, at age 95," he told me. Like his son and his grandson, Dr. Henry Lax was "a charmer", and his charm did not exclude the fair sex.

Anneli Cahn

To describe Peter's family life, I start out by introducing Anneli Cahn, his future wife, whom he met in the complex variables class at NYU. We will look at her mathematical career in Appendix 1. Anneli was born in Katowice, then a German city, now part of Poland, on February 23, 1922. Her education was repeatedly disrupted by her family's fleeing from the Nazis. They moved from Katowice to Berlin, from Berlin to Paris, back to Berlin, and then to Tel Aviv, finally immigrating to New York. When Anneli studied Euclidean geometry in the lyceum in Berlin around 1935, she was attracted to the constructions and the logic, and she easily solved advanced problems. "I didn't have to look up anything; I didn't have to consult libraries or books. I could just sit there and figure things out."

Her family settled in New York in 1935. In a high school in Queens, she studied geometry with Miss Eaton, "a lovely old lady." The algebra class that followed was a disappointment: "It looked like a lot of tricks. It was only much later that I saw that there was a foundation to the stuff... I never liked rules that didn't have a basis that I could understand."

Anneli earned a bachelor's degree from Adelphi University in 1942 and moved on to graduate work at New York University. She took the graduate course in complex variables in the spring of 1943, and there she met Peter Lax, who was still an undergraduate.

He says, "I had no problem picking her out. Anneli was extremely beautiful, by far the most beautiful girl in the class. She hasn't changed much, her beauty is in her bone structure. We got acquainted by the time-honored method of the male showing off. The peacock has his tail, and I had my mathematics."

The complex variable instructor was a holdover from earlier times at NYU, before Courant came. Professor Putnam taught from his notes of Courant's lectures. "Whenever someone asked a question, he became confused. I already knew complex variables; I had read Knopp. After a while Putnam noticed that I always knew the answer, so he began calling on me to answer questions. He realized that I knew more of the subject than he, so he let me just sort of take over the class."

This impressed Anneli.

"But I didn't date her then. I was only seventeen, and she was married—a marriage she entered into mainly to get out of her parents' house."

Anneli Cahn Lax

Anneli: It was clear that he was a mathematician. It wasn't clear whether he was socially awkward or just putting on an act. There was this rather charming mixture of mathematical know-how and shyness, and quite a bit of wit. That marriage was very brief and was later annulled. I was no longer married by the time Peter had his first furlough from the army.

Peter: After I was drafted, I didn't meet Anneli again until I returned from the army in 1946. My first problem was to get a bachelor's degree, which I managed by scrounging together credits from NYU, Stanford, Texas A&M, and the University of New Mexico. I got my A.B. in February 1947.

While Peter had been in the army Anneli dated several other young mathematicians. But once he was back, "It didn't take me too long to fall in love with Anneli. I maneuvered to get into the office in the Bible House that she and Cathleen shared."

Cathleen Morawetz writes, "I first met Peter before he was discharged from the army. He came to lunch in his uniform and he joined the usual crowd at the Eighth Street Delicatessen, where we often ate. He made a youthful and enthusiastic

Cathleen Morawetz, 1983

impression. We all moved to the 'Bible House' at Astor Place in the fall of '46. There Anneli, Peter and I shared an office. Most of the male students did not want to be in an office with women, but Peter was even enthusiastic about it or at least appeared to be. While we were there, Courant was giving a course on Dirichlet's Principle, but he was not there for a good many lectures and Peter gave the lectures instead. Usually Peter had no advance notice."

Cathleen Morawetz, the daughter of the well-known Irish-Canadian applied mathematician John Synge and wife of a well-known polymer chemist, was my first boss when I became a teaching assistant at NYU. She had earlier been the key link between Richard Courant and Kurt Otto Friedrichs in their hard-fought collaboration on the classic treatise *Supersonic Flow and Shock Waves*. Her own research included famous breakthroughs on "transonic flows", compressible flows that are "breaking the sound barrier." She has been the president of the American Mathematical Society and holds the National Science Medal. To this day she is a close friend, even though we rarely see each other any more.

Peter and Anneli got married in 1948; he got his Ph.D. in 1949.

Anneli was interested in languages and wanted to learn Hungarian for fun, but Peter said no. "I did not want a Hungarian family. I wanted an American family. I don't think I ever forgave the Hungarians for their treatment of Jews during the war...."

In 1950 and 1954 Anneli gave birth to sons John and James.

Peter was still able to concentrate on his work. Having children around didn't interfere. He said, "I was sufficiently single minded about mathematics that having children didn't hamper me. But you will have to talk to Jimmy about that."

Peter adored his two young sons. But as he became more in demand all over the world, he was often away.

John and James

Anneli was a very conscientious, involved mother, but she had a demanding, full-time job at the university. And she naturally loved solitude. She spent whole summers at Loon Lake in the Adirondack Mountains, even when Peter was unable to be with her.

Anneli said, "He can think about a problem and still talk to you or read a book. There's always something going on. In my generation, when a woman got involved with a family, it was assumed that she took responsibility for the household and the children. It would be important for her to have a partner, not so much to help, but one who's 'with it' in spirit. Peter could seem to be 'with it', but with his brilliant mind, he was not always 'with it'. He knew very well how to protect himself from distraction. He had a whole separate life that I did not share, and that was very hard."

Like Peter's parents, Henrik and Klara, Peter and Anneli hired a nanny to help with the children.

Jimmy

Peter's younger son, Jimmy, like his grandfather, is a successful internist practicing in Manhattan. He lives in The Eldorado in the apartment that used to belong to his grandparents. Jimmy married another physician, Jocelyn Brown, and they had three fine young sons. But then an irreparable chasm opened up, and they divorced.

Jimmy's sons traveled back and forth between Jimmy's home and their mother's. Tommy was a curator in the Harlem Studio Museum. Timmy and Pierre were finishing college.

To see Peter with his son and his grandsons is to see very visible expressions of doting affection. It is really heart-warming, in contrast with the indifference or hostility that are so common between fathers and sons. In this respect Peter is continuing a relationship that he had with his own father. He has told me how affected he was by realizing, in his teens, that "my father would do anything for me."

Jocelyn Brown was the daughter of an African-American man who studied medicine in France and married and had a family there. So Peter's three grandsons are ethnically blended—Hungarian-Jewish-Franco-German-African-American. Peter's close connection with them helps him to embrace their generation's beautiful diversity.

One July afternoon while I was back East, I took a day to go to New York to work on this book. I made an appointment to meet Jimmy at his apartment on a Wednesday evening and asked him to have his sons there. When I got to Jimmy's apartment in The Eldorado he was not at home, but a friend greeted me, a lady named Dee Dee who is in her nineties and very perky. She was his patient and became his friend. Quite talkative and cheerful. Eventually Jimmy showed up; he bicycles from his office and had to stop off at a hospital. Two of his sons, Tommy and Timmy, were there, and someone brought a lot of Chinese takeout for dinner. I knew Tommy from when my partner Veronka and I had stayed in Peter's apartment a few years ago for Veronka's granddaughter's wedding. Now Tommy has become a museum curator. Timmy is uncertain of his future; he loves writing and very much enjoys working at a farmer's market during the summer. He looked tanned and healthy. The other son, Pierre, was away working at a job on Martha's Vineyard. I had dinner with Jimmy, Tommy, Timmy, and Dee Dee.

The rest of this chapter is based on my voice recording at that dinner and at a subsequent visit to Dr. Marnie Greenwood, who was the fiancée of Peter's older son, John.

Jimmy: Tommy was interested in math when he was really little. I remember once Anneli or Peter told him about square roots, and then one of their friend mathematicians said to Tommy, "Well, if you like square roots, let me tell you about cube roots" and they were off! [*chuckling*] He was very little then.

Reuben: With grandchildren you have all of the pleasure and none of the pain. But now walking is painful to Peter. It's tough. I tried to convince him that there is nothing wrong with using a wheelchair.

Jimmy: Don't use that word in front of him. [*laughter*]

Dee Dee: Or a cane. Canes are out too, honey. You won't be his friend. [*laughter*]

Jimmy: Just block it out.

Reuben: I don't expect you to repeat everything I say.

Jimmy: And likewise.

Tommy: I thought it was just a swagger.

(This conversation took place before Peter had knee replacement surgery. He is now walking comfortably without pain.)

Reuben: How good a tennis player is Peter actually?

Jimmy: He's pretty good. But not the greatest. Anneli was a brilliant figure skater. I have a movie of her where she's twirling. And she was on the swim team.

Tommy: Peter was very competitive. But when we asked Peter, "Are you competitive?" he'd say, "No, no."

Jimmy: Peter plays Ping-Pong too.

Reuben: He told me you have a Ping-Pong coach.
Jimmy: I do.
Reuben: I've never heard of that. That's unique.
Jimmy: In fact his aunt, my grandmother's sister, whom I knew quite well, was the Ping-Pong champion of Hungary in 1932. So we've always played Ping-Pong in this family. At the end of her life she lived in New York and Chicago. She lived with us for a while, with my grandfather.
Reuben: The last time Peter visited me in Santa Fe, we tried to play a little Ping-Pong. He hit the ball off the table a couple of times, and then explained that he needed a heavier paddle. He said he'd bring his paddle with him next time.
Jimmy: He's a little competitive. I'm much more sympathetic to my parents since I've been a parent and had children. They were both such great grandparents to my children. They helped me out. They were so much more available as grandparents than as parents.
Reuben: As parents they were not so available. What's it been like to be your father's son?
Jimmy: He was a very young father. He traveled a lot. And my mother sort of had a short attention span. When you were with her and she was giving you attention, she would concentrate for a while, but then it was, "For now I'll give you attention, but what's next?" Unfortunately, I've inherited that. Anneli was not like my father with his Mt. Olympus-like mind. She had to really work to get her Ph.D. and tenure. Plus the responsibilities of the household fell on her shoulders.

"But we had a wonderful nanny, Elizabeth Thomas. She was with us for about 22 1/2 years, from when I was five. At the end, she would just come in and sort of hang out. She wasn't working much anymore. She was a member of the family. She was a mother figure for me, sort of another mother. She was African American, born in Augusta, Georgia. Not much education, but you would never know it. She had incredible diction. She was a good psychologist. She was very smart, very in tune to children."

Reuben: Is she still alive?
Jimmy: No. She died just six weeks before my brother.

Jimmy and Johnny attended a famous private school, the Fieldston School, in Riverdale, way up in the Bronx. It is part of the Ethical Culture Society.

Both boys had difficulty learning to read. Dyslexia was then not so commonly talked of and well known as it later became. However, in the course of time, with help and support, both boys became academically successful.

Reuben: How old were you when someone told you Peter was famous?
Jimmy: It was probably a blessing that I grew up unaware that my father was famous. It would have been very intimidating, I think. I didn't know what my parents did. Occasionally they would speak math. It was like another language. Like when they didn't want me to understand and spoke German. They would talk math at the dinner table, and I would tune out.

"But my mother was very good at helping me with math and explaining when I didn't understand something. If I couldn't solve a problem, rather than making me feel I failed, she could understand where I got tripped up. And she spent time to do that. That was great.

"I had learning issues, like dyslexia. I didn't learn to use textbooks until I went to college. Before that, I would learn everything by listening and then asking my parents to explain. [*laughs*] My mother would look at my textbooks and then explain things. I didn't really understand that you could learn something by reading until I was in college.

"One of the best things about my parents was that they had nice interesting friends who became friends with me and my brother and are my friends to this day. When Peter and Anneli talked math it could be numbingly boring, but it could also be fun to watch them talk about math problems so enthusiastically. When we were kids sometimes a mathematician would come up to us and test us to see if we were prodigies. I would fail the test really quickly, so they would move on. [*more chuckling*] My father was a prodigy, so they figured Johnny and I had a double dose of math genes.

"Anneli went to college when she was sixteen and graduated when she was nineteen. She loved languages; they came easily to her. But mathematics was more challenging; she didn't get it so easily, so she found it much more exciting. She decided to challenge herself and do it, and she enrolled at Courant. It's called masochism. She was a little perverse, in a way. She didn't care what other people valued; she was almost like an anti-snob. If society thought that you should do 'x' she would do 'y'. In fact, she didn't like awards and prizes. She thought your enjoyment of your work should be your reward."

Reuben: You don't agree?

Jimmy: Oh, I do agree. But I think it should be balanced: it's nice to get recognition. Fortunately, not everyone has to get an Abel.

Reuben: If everyone got it, it wouldn't be worth anything. So Anneli had a lot of guts and a lot of courage, and she was there for her kids....

Jimmy: Luckily, unlike the situation with my brother, I did get to resolve some of the issues with my parents, and ultimately Anneli was there for me.

Reuben: How did you resolve it?

Jimmy: By talking to her. And by trying to understand. As I mentioned earlier, it helped for me to be a parent, to help my understanding of where they were....

Reuben: You could do that?

Jimmy: I'm just that kind of guy. I can talk about things, and how she loved me.

Dee Dee: That's why he's such a good doctor. He shows that with his patients too.

(It was Dee Dee who had answered when I rang the doorbell, expecting Jimmy. He was delayed and had asked her to let me in. Before he got home she told me that she had first met him as his patient before becoming his friend. I took her for seventyish, but she told me she was ninety-seven years old.)

Reuben: Well, that's what you say, Dee Dee, but what about all the other patients?

Dee Dee: Oh, they love him. I know.

Reuben: That's not biased?

Dee Dee: Oh, not at all. It's true. I'm often sitting there in his waiting room and there are other people sitting there and they'll say, "Isn't he wonderful?" And nobody asked them; they just say it.

Johnny

Johnny was "a handful", as they say. Charming and attractive, with a strong physical resemblance to his father, he was often overenergetic and hard to control. Peter said, "Every time I went to the school, I found him in the corridor outside the classroom." The school labeled him as ADD (attention-deficit disorder) and recommended treatment: dosage with Ritalin. The personality change that the drug produced was not necessarily what Anneli and Peter really wanted. Peter couldn't stand it; he insisted on stopping the Ritalin doses. "I threw them away," he says. "It didn't make sense. I liked him the way he was." In the third grade Johnny started to get better.

One aspect of Johnny's wild or mischievous side was the way he sometimes treated his kid brother, Jimmy. The older brother is naturally "the boss" of the two and can be a model for a younger brother. But Johnny seemed to enjoy teasing and mocking his kid brother beyond what Jimmy could enjoy or even endure. Eventually, Johnny's fiance, Marnie, said to Johnny, "Why are you so shitty to your brother?" and then things changed for the better.

Jimmy: He was a tough older brother. He was a very charming guy with a lot of warmth, and we were very close, but he'd also be competitive, sadistic, and difficult. He had really bad ADD. We had a very intense relationship, and we did talk about some of these things, but unfortunately we didn't have time to resolve them.

Reuben: Did your parents deal with this problem? Were they aware of it?

Jimmy: If they had been more present, the problems wouldn't have been so bad. I think that their absence physically and emotionally was difficult for my brother and for me

Reuben: How did you respond to the pressure?

Jimmy: Well, Elizabeth was great. My defender. You do have to stick up for yourself, but Johnny was unpredictable. Sometimes he was the good Johnny and sometimes he was the bad Johnny. And you didn't know which it was going to be. I was just talking to Jill about that, and those were her words. Jill (Abramson) is my childhood friend; we met in the sandbox. Now she's the illustrious executive editor of the *New York Times*. She knew Johnny very well. One of the tragedies of my brother's death is that he and I hadn't worked out all of our issues.

After surviving turbulent school days in elementary school and high school, Johnny evolved into a successful college student at Brown University. He studied American history and became a protégé of the historian Abbott Gleason. Gleason wrote a touching memoir of Johnny in his autobiography, *A Liberal Education* (Appendix 7, below). Johnny's undergraduate thesis was about black jazz musicians in Chicago in the 1920s. It was published in the *Journal of Jazz Studies* (Appendix 8).

John liked to show up without introduction at departments of mathematics and demand an answer to the question, "Who is my father?" People could easily guess the answer: Peter Lax. He became a history graduate student at Columbia University, taught history at Mount Holyoke, and was working on a dissertation about the history of the American Legion. That was the main organization of U.S. veterans of World War I. It played an active, sometimes reactionary, role in American politics in the 1920s and 1930s, and again in the 1950s. John's research was used in a book by his friend Bill Pencak.

In 1982 he was driving through Chicago on his way to the Herbert Hoover Presidential Library in Iowa and was struck by a drunken driver. He was killed instantly.

There is no way I can describe the blow to Peter and Anneli from the shocking death of their son.

I happened to be visiting the Courant Institute shortly afterward and went to Peter's office. Anneli was there. She asked me to go sit with Peter. I did so. Peter said only, "It is very hard." There was nothing to be said by me.

As a memorial, Anneli and Peter created an annual lecture at Mount Holyoke College to be given by a historian of the highest distinction to commemorate the work and spirit of John Lax.

Johnny's fiance, Dr. Marnie Greenwood, became a close friend of Anneli, Jimmy, and Peter. She continues to be a member of the Lax family. I met her at her home near Hartford, Connecticut.

Marnie: I first met John in Brookline, Mass., on New Year's Eve. My roommate Jeanie and I were sitting at home. I wasn't going out with anyone, and my roommate Jeanie had just broken up with someone. A friend of Jeanie's called and asked, "What are you doing?"

"Jeanie said, 'We're just hanging out in our apartment.'

"They said, 'Can we come over?' and John was with them. He was wearing a fur coat, of all things. It belonged to John's grandfather, Henry. A patient of Henry's had given him this fur coat. I will never forget how John walked in that door. He had that look, that characteristic look that I remember so much. That classic John Lax look, the way he'd look up at me. It was just this shy smile. It was engaging. He was this incredibly engaging man.

"After John and I had been corresponding for a while, we went up to Loon Lake for a ski trip. But he wasn't that experienced a driver. During the night he drove off the road onto the median!

"So then we had to hitchhike. The police drove us to a motel. That night we stayed in a motel. The next day a tow truck came to tow us in. It was really tight in the front seat with both of us next to the driver, so John puts his arm around me and says to the truck driver, 'It's amazing what I'll do to get my arm around this girl.'

"The cabin was barely winterized. We had no food. There was no grocery store out in the boondocks. But the Mosers happened to be there. They fed us. They took us skiing. I like Gertrude Moser; she's a very real, straightforward person."

Gertrude Moser is the older daughter of Richard and Nina Courant. She was married to the mathematician Jürgen Moser.

Reuben: Is Gertrude like her sister, Lori?

Marnie: They're different. Lori is such a warm person, so easy to talk to. Gertrude is more reserved. She isn't one to initiate a conversation.

"After that weekend, it worked out a lot better than I thought it was going to. The following year John and I moved in together. He was in graduate school, working on his thesis. For almost three years he was traveling back and forth between Cambridge and Chicago."

Reuben: When the accident happened in Chicago, was somebody driving down the wrong side of the road?

Marnie: On a highway the other driver veered across the median into oncoming traffic. John was just driving along...

At John's memorial service, three speakers were Thomas Bryson, Abbott Gleason, and William Leuchtenburg. Bryson said, "To be John's friend was to be engaged with a life force of the first magnitude. His intelligence, kindness, wit, charm, and infectious enthusiasm for a range of interests penetrated the lives of everyone he knew. I brought him home to family gatherings. He would talk sports with my father, arm-wrestle my brother, debate the meaning of romantic love with my mother, swap macho stories and jokes with my uncles, and gorge himself on the available food. All of this within the first half hour of his arrival."

Gleason said, "The hours spent talking with John about everything under the sun, the sense of the possibilities of friendship which knowing him has given me—these have deepened me in a way that I still don't fully understand. What wisdom I have, what capacity for good offices, what reserves of benevolence for my students, and above all my sense that with our friends we must seize the day have all been enhanced by John Lax. He became part of me."

Leuchtenburg said, "What I noticed most of all was his smile—at first shy and a little doubtful, but then just covering his whole face with a broad grin. I cannot bring myself to tell you how awful is my sense of loss. These final words sum up, better than anything else, what he felt about the world: 'I would be impossibly myopic, egotistical and narrow if I did not realize that if anybody in this world has nothing to kick about it is I. I can't think of anybody off the top of my head who has been treated with more kindness, consideration and love, as well as being exposed to many first-rate minds from an early age...this wonderful good fortune that has seemingly been with me all my life.'"

Marnie said to me, "John adored his parents. They were such lovely people, he put them on a pedestal. They came over to visit all the time. Anneli was such a generous and giving person. With Jimmy...how much she wanted him to be happy! She was the one who was so empathetic to Jimmy. John could be relentless in kidding someone, going on and on. But Jimmy would get hurt by it. Jimmy's a sensitive person."

Reuben: Peter felt you changed all that...

Marnie: Maybe I did. Sometimes you never know the impact you have on someone's life. And I think this had a big impact on Jimmy. He'd get angry at John; he was quite young at the time, and I got to know Jimmy a little bit separately from John. I got to develop my own relationship with Jimmy. I consider him my brother.... I decided to go back to school to complete the requirements for medical school. Anneli became a mentor to me. I was a first-year medical student when John died. After John died, they were really great to me. I never felt that I was a substitute for John. I felt that they, like most of my best friends, are friends John and I knew. Anneli was very open. I remember her saying, "You and John were so capable of having a wonderful relationship. I hope you realize you will go on to have another one." Peter doesn't talk about himself or how painful this was.

"Then we all went to Europe. I arrived in France before Peter, and Peter's father, Henry, was already there. He picked me up in southern France. I remember meeting this impressive man, and I was sort of a little shy.... We did a little traveling in Switzerland. He said, 'You need to eat more, my child.' [*laughs*]"

Peter's father objected to Marnie's going into radiology, which he considered a dull specialty. But Peter explained that with modern advances, it has become interesting and demanding.

In July and August every year Peter is in the Adirondack Mountains of northern New York, not far from Canada. His place there, called Prince Camp, is next to Loon Lake. The name "Prince Camp" is inherited from its first owners, the Prince family of New York. In the late nineteenth century, this region was popular with rich New Yorkers. They would stay the whole summer. It was far from the city and close to the racetrack at Saratoga Springs. The Prince family built a structure commodious enough for themselves and their guests, cooks, chauffeurs, nannies, and what not.

Marnie told me, "After John's death, I continued to go up to Prince Camp. Jimmy and I have gone with our children for many years and continue to spend a week together up at Prince Camp every year." She and her husband, David, were leaving for Loon Lake in another two weeks.

Reuben: How do you get there?

Marnie: With difficulty. We go for a week. The number of people that come sort of goes like this: Jimmy's three boys are going, and they all bring either significant others or friends. So already that's six right there. My daughter's coming. Jimmy's bringing a really dear elderly friend of his, Dee Dee. And Jimmy's bringing another friend, and then John Lax's best friend from college is coming. So all I can think about is all of the cooking and cleaning.

Reuben: How do they manage? Shopping and cooking and all that? How is it done?

Marnie's husband, David, answered, "Jimmy usually stops to shop at Plattsburg. He likes to cook, as do I."

Reuben: You cook for twenty people?

David: Yeah. Actually, mostly the dinner and some breakfast stuff. Then people help themselves.

Reuben: So it's all set up?

Marnie: It's a big kitchen.

David: It's shabby chic. There's a lot of stuff there.

Reuben: What do you do while you're there besides cook?

David: They all swim constantly and fish.

Marnie: David fishes, and if he can [he'll] hook a kid to go with him. No one else fishes.

Reuben: Isn't it cold? It's too damn cold up there.

Marnie: We go in no matter what.

Lori

Peter and Anneli started going to Loon Lake in the summer in the 1960s because Anneli's friend Kitty Sidrane went there. In 1972 they heard a rumor that Prince Camp might be "going on the market". Peter immediately jumped into a canoe and rowed across the lake to the Prince Camp boathouse. The owners said, "We decided to sell only two days ago, how did you know already?" It turned out that their caretaker's son was friends with the Sidranes' sons. The owners of Prince Camp were asking only $135,000, which Peter immediately agreed to. It was the very first offer the owners had received. The Sidranes split the price and are still partners. Today the camp is worth $1 or $2 million.

The camp is far from modern, but it has great historical interest and plenty of room for guests and friends. Peter has counted twenty-six beds! The lake water is very cold, but clear and clean. The cry of the loon is occasionally heard.

As time passed, Peter and Anneli were joined in their Loon Lake summers by two more Courant Institute couples—the two daughters of Richard Courant, Gertrude and Lori, and their mathematician husbands, Jürgen Moser and Jerry Berkowitz. Moser was a German mathematician famous for work on dynamical systems, among other things. Berkowitz was a brilliant NYU math student who had become a faculty member.

After Anneli Lax and Jerry Berkowitz died, Lori and Peter became each other's second spouse. Peter Lax and Lori Courant first met in 1942, when she was thirteen and he was sixteen. Her father, Richard, brought Peter home to New Rochelle to meet the mother, Nina, the two brothers, Ernst and Hans, and the two sisters, Gertrude and Lori. Peter was very shy at first and didn't speak much English.

Lori was then learning violin and viola from Nina and other teachers. Nina sometimes let Lori use her precious violin. Gertrude, the older sister, was witty and funny like her father but not as pretty as Lori. It is said that once, when Gertrude ran away from home and was gone for a whole day, Nina never even noticed.

Lori was greatly hurt by a strange incident. One summer, Richard Courant planned a family trip across the U.S. by car. At the very last minute, he told Lori that she wouldn't be coming. He said there was no room for her in the car. She says he simply didn't want her.

Ernst and Hans became physicists, Ernst a distinguished one. Gertrude studied biology and did publish one paper, but then disappointed her father by giving up science. Lori followed Hans and Gertrude to Swarthmore College. Richard did not permit her to choose a music conservatory, nor to major in music, because music is unwise as a career choice. She also loves to write, so she majored in English. After college, she studied full-time for four years at the Manhattan School of Music and earned a master's degree with the famous teacher Raphael Bronstein. Bronstein was a pupil of Leopold Auer in St. Petersburg and wrote the book *The Science of Violin Playing*. Lori performed for years with the American Symphony Orchestra under Leopold Stokowski, where she was first-desk violinist. (She says that Stokowski was not Russian but an Englishman named Stokes. His Russian accent varied a lot.) She taught viola, violin, and theory at the Manhattan School. She was principal viola in the Grand Teton Music Festival and Symphony of the New World. She appeared with the Parnassus Chamber Ensemble, Anacrusis, the Piatigorsky Foundation, and Musicians for Peace, and several times with the Albert Einstein Symphony. She is very active to this day in chamber music and teaching.

Anneli Cahn, later to be Anneli Lax, was a favorite of Papa Courant. Early in her marriage to Peter, she took care of the Courant's New Rochelle house while the Courants traveled in Europe. An important romantic incident occurred when Peter and Anneli had tickets to a chamber music concert at Washington Irving High School, along with their friend Jerry Berkowitz, a brilliant graduate student at the Courant Institute. The evening of the concert, Anneli had to keep another appointment, and Peter suggested giving Anneli's ticket to Lori Courant. Lori had

Lori Berkowitz playing her "fiddle"

already met Jerry, but that concert evening brought them closer, and eventually they married. Peter still takes pride in his role as matchmaker. Jerry and Lori moved into a shabby apartment on the Lower East Side of Manhattan. Later they moved into a somewhat larger shabby apartment on Second Avenue.

Jerry became a faculty member at Courant. He did not become an outstanding research mathematician, but I greatly enjoyed and appreciated his teaching when I took his class in functional analysis in 1959. The two Courant sisters and their mathematician husbands joined Peter and Anneli as owners of vacation homes at Loon Lake.

Lori and Jerry had no children but eventually adopted a son and a daughter, David and Susan. David is in business, and Susie runs a restaurant in Lake Placid, a popular resort town in the Adirondacks. Jerry suffered from diabetes, and he wasn't as rigorous as he might have been about checking his glucose levels. He had lost some weight, but was still a little overweight. It was a complete shock when he died suddenly of a heart attack in 1998 at age seventy.

The following year, Anneli Lax died after two years of struggling with pancreatic cancer diagnosed by her son Jimmy. Peter and Lori both continued to go to Loon Lake in the summers, but now alone, without partners. I actually visited Peter there for a few days. I enjoyed the hiking more than the ice-cold swimming. One evening the two of us were invited to supper by Lori Berkowitz. It was the first time I met her, although years before, while a graduate student, I had done copy editing for her father. As time went on, Lori and Peter found each other's company and support indispensable, and in 2006 they married. After a while, Lori gave up the Second Avenue apartment. While working on this book, I have enjoyed a couple of long visits with Peter and Lori on the ninth floor of The Eldorado.

Jerry Berkowitz working with Harold Weitzner

It is a rewarding experience just to watch Peter and Lori walking down the sidewalk, arm in arm. Years of tennis had left Peter with painfully worn-out knees, but since his knee replacements he is walking easily and playing tennis again.

CHAPTER 4

Early career

Soon after the Laxes arrived in New York, in the Spring of 1942, Peter's father took him to see Courant. "I was not yet sixteen. I remember very well—Courant got very excited, and we talked about mathematics. But I had to finish high school, so the first concrete thing that developed from that meeting was that my father became Courant's doctor." Richard Courant had been hired in 1934 to develop NYU's mathematics graduate program. He structured a new institute with no separation between pure mathematics and applied mathematics.

"Courant had very original and very unusual ideas of collaboration," says Peter. "Mathematics for most people is a very lonely profession. Of course there are collaborations, but it is typically the lonely individual thinking about the problem. That was not Courant's idea.

"Friedrichs's coming to the institute in 1936 was very significant. Friedrichs was not Jewish, so he was allowed to keep his position at Braunschweig, but he became increasingly unhappy living under the Nazis. He wrote to Courant in 1935 that he is ready to come to the U.S. if he could get a professorship. Courant wrote back that such positions were not easy to get, but that he will try. Soon after, Friedrichs wrote again that he is willing to take any job except dishwasher. And shortly after that came a third letter. 'Willing to wash dishes.'

"Stoker's path to NYU was equally interesting. He was an engineer specializing in mechanics. In 1935 his wife unexpectedly inherited $10,000; they used that money to go to Zürich for two years, where Stoker wanted to study with a distinguished professor of mechanics. The first course he took was on geometry, given by the distinguished geometer Heinz Hopf. Stoker fell in love with the subject and got a doctorate in mathematics. Hopf had been a close friend of Courant and wrote to him recommending the "junger Amerikaner" as someone who could help Courant build an institute dedicated to both pure and applied mathematics. They formed a triumvirate. And they had brilliant students."

The Japanese raid on Pearl Harbor in December 1941 brought the U.S. into World War II. "Suddenly there was federal money available for research. That was what turned a very modest enterprise into a much bigger one. It became quite clear to the government what Courant had known for some time—how important technology and mathematics and science were to national defense, as exemplified by radar and the atomic bomb and supersonic aircraft.

"America's entry into the Second World War transformed most American academic scientific institutions, none more than Courant's operation at NYU. Government funding was made available for research relevant to war work through the Office of Scientific Research and Development (OSRD). Its head, Vannevar Bush,

Louis Nirenberg

Courtesy of Louis Nirenberg

saw the importance of mathematics for the war effort and set up the Applied Mathematics Panel under the direction of Warren Weaver. Courant was soon invited to be a member of this elite group."

The mathematical project at NYU sponsored by the OSRD was about the flow of gases in general and the formation and propagation of shock waves in particular. There was enough money to support young research associates (Max Shiffman, Bernard Friedman, and Rudolf Lüneburg) who also served as adjunct faculty in the graduate school. There was also money to provide stipends for graduate students, some of whom were drawn into war work. Courant insisted that graduate training continue even during the war.

Serendipity brought Louis Nirenberg to the Courant Institute. In the summer of 1945, he was doing atom bomb research at the National Research Council of Canada in Montreal. Two of his colleagues were Ernst Courant, Courant's oldest son, and Ernst's wife, Sara. "She's the one who led me to Courant," said Nirenberg. "It was pure luck. I came down in the summer of '45 to be interviewed by Courant and Friedrichs. They agreed to give me an assistantship, so I came back in the fall. At the time, the offices were in Judson Hall, on the south side of Washington Square Park. There were some young assistant professors, Max Shiffman and Bernard Friedman, and just a very few of us students: Eugene Isaacson, Martin Kruskal, Peter and Anneli Lax, Joe and Herb Keller, Cathleen Morawetz, and Harold Grad. It was a small group of graduate students but a very bright bunch of people. Courant's idea was unlike most mathematics departments. Typically, after people graduate with a Ph.D., they're sent to other universities, but Courant's idea

Peter Lax and Louis Nirenberg at Professor Nirenberg's apartment on New York City's West Side

was to keep the very good students. He kept a whole bunch of us for a number of years. It was a great time."

It is truly amazing that of the eight or nine graduate students with Richard Courant in the mid 1940s, virtually every one would become a full professor at NYU and an internationally recognized authority in his/her specialty.

The advent of the computer

Peter said, "The connection between computing and mathematics can be traced back to Courant, the Courant-Friedrichs-Lewy paper. Friedrichs himself, somewhat strangely, was not interested in actually doing computing. Courant was, but perhaps he was too old to do research by the time the big machines came in, but his spirit remained." Richard Courant had been quick to recognize the crucial importance of computing machines for the future of mathematics research and teaching. Two "super computers" served NYU through the fifties, sixties, and early seventies—first the UNIVAC and then the CDC 6600.

When the University of California opened a new research campus at La Jolla near San Diego in 1950, mathematicians at UC hoped that the new campus would concentrate on applied mathematics and that Peter Lax would become its director. There was a sustained effort by UC to attract Peter, but he never seriously considered leaving New York and the Courant Institute.

In 1953 the Atomic Energy Commission's much-sought-after UNIVAC supercomputer contract was up for grabs. Richard Courant's tenacity, his Washington connections, and his fighter's will made sure it was awarded to NYU.

Robert Richtmyer (1910-2003)

Remington Rand UNIVACs were the first big electronic computers commercially available in the U.S. The first one went to the Census Bureau, the second and third went to the air force and the army, and the next two went to the Atomic Energy Commission, which leased one each to NYU and Livermore Lab in California. This victory was a major turning point for the institute. It guaranteed federal investment in NYU's math department. The UNIVAC I was first priced at only $159,000, but ultimately sold for between $1,250,000 and $1,500,000. Forty-six UNIVACs were built and delivered.

Peter remembered the UNIVAC as having a memory of 1,000 words; the memory was 100 channels of 10-word mercury delay line registers. A word was 11 decimal digits, plus sign. The machine took up 35 1/2 square meters of floor space, weighed 29,000 pounds, used 5,200 vacuum tubes, and did about 1,905 operations a second "which is nothing today. Then it was huge." It was inaugurated at a ceremony at NYU in 1953. Participating were James Rand, the CEO of Remington Rand, and his whole board of directors, including the chairman, General Douglas MacArthur, and board member General Leslie Groves, head of the Manhattan Project. A photograph shows Generals Groves and MacArthur with Richard Courant. Peter is there too.

To direct the computing center, Courant hired physicist Bob Richtmyer, the head of the theoretical division at Los Alamos. While at NYU, Richtmyer wrote the influential *Initial-Value Problems for Partial Differential Equations*. Its second edition, which he co-authored with K. W. Morton, includes my own thesis results on mixed problems.

Peter's best known contribution to numerical analysis, the Lax equivalence principle, was published in 1956 in a joint paper with Bob Richtmyer. Peter wanted to correct a misapprehension that was then widespread among people doing computing work. To solve differential equations, one commonly approximates them with "difference equations", where the approximating solution takes small finite steps over a "mesh". Two crucial issues are stability and convergence. A difference scheme is said to be stable if as the mesh size is made smaller and smaller, the associated solutions don't "blow up". And it is said to be convergent if the solutions then become arbitrarily close to the actual solutions of the equation which the difference scheme is approximating.

There is also an issue of "round-off error", because differential equations use infinite decimals, whereas computers must "round off" to a few binary places, usually 8 or 16. The resulting "rounding errors" can pile up, even though one expects them to be random—equally likely to increase or diminish the results.

"In a lecture I heard at Los Alamos, the speaker asserted—and I suspected he was not alone in his belief—that the blowup of unstable schemes is due to the magnification of round-off errors. Then if you had a machine (which of course you never could have) which calculated with infinite accuracy, then even a nonstable scheme would converge! I wanted to explode that idea."

Before the Lax-Richtmyer paper of 1953, analysts thought they had to prove both stability and convergence. Stability, because if the calculated solutions magnify round-off errors as the mesh is refined, they are useless. Convergence, because the whole point of the difference scheme is to approximate closely the actual solution of the exact differential equation.

The equivalence principle states that stability plus consistency is equivalent to convergence. In other words, a numerical approximation scheme is convergent if and only if it is both stable and consistent.

It was already well understood that if a scheme is consistent and stable, it must converge. But it was thought by some that convergence was possible even in the absence of stability if only the round-off error was controlled. This was the fallacy that Peter wanted to correct. And he did so by proving a rigorous theorem. That is the main part of the Lax equivalence principle.

The proof is quite short and simple. It's an application of a basic theorem of functional analysis, the principle of uniform boundedness, or its equivalent, the closed graph theorem. (See Appendix 4.) At that time, few mathematicians would have been familiar with both the closed graph theorem from functional analysis and the issues of stability and convergence from practical computing. Peter's forte is to solve practical problems using ideas from advanced pure mathematics.

The closed graph theorem says, roughly, that a closed linear mapping is continuous. Peter Lax was in a position to notice that the desired stability of the difference scheme is an example of the abstract condition of "closedness" and the desired convergence is an example of the abstract notion of "continuity". The closed graph theorem then concludes that a consistent finite difference scheme is convergent *if and only if* it is stable. The proof takes just a page or so, but it could only be discovered by a mathematician with feet in both camps.

In 1957 Peter published another influential paper, "Asymptotic solutions of oscillatory initial value problems". Judging by the title, it seems to be about a rather specialized classical topic, but it contains a very powerful general idea, which, in the hands of Lars Hörmander and his followers, became a popular, productive approach to partial differential equations under the name of "Fourier integral operator calculus". Asked to explain what made this paper so important, Peter answered, "It is a micro-local description of what is going on. It combines looking at the problem in the large and in the small, and that gives it its strength. The numerical implementation of the micro-local point of view is by wavelets and similar approaches, which are very powerful numerically."

The paper was partly motivated by earlier work of Morris Kline and Joe Keller (with collaborators) on the problem of diffraction. Diffraction is a property of all wave motion. In sound waves, diffraction permits one to hear a person behind a tree or in the next room. It was the recognition of diffraction effects that established the wave nature of light, in opposition to Newton's particle theory. One aspect of diffraction is interference, the possibility for two waves out of phase with each other to cancel rather than add. In quantum mechanics this is the two-slit experiment, which shows the wave nature of elementary particles. Mathematically, diffraction effects are of second or higher order with respect to the frequency of the wave. The first-order term corresponds to the particle-like behavior of the wave, which is called "geometrical optics". In a formal, nonrigorous way, geometrical optics and diffraction had already been studied by physicists and applied mathematicians. Starting from Maxwell's equations of electromagnetism, they made what is called an "asymptotic expansion" with respect to the frequency of the wave. Peter ingeniously connected the asymptotic expansion with the rigorous theory of characteristic curves and surfaces. Thus the asymptotic expansion of formal diffraction theory was made rigorous. This rigorous theory even worked in the much more difficult case when the medium is not "homogeneous", that is to say, when the coefficients in the wave equation are not constant but vary with position in space. That is what Peter means when he says his method is "micro-local": it combines asymptotic aspects in both the frequency domain and the spatial domain.

"Fifteen years after this paper was written, Lars Hörmander and Hans Duistermaat microlocalized the approach developed here" (V. Guillemin, p. 119, Lax, *Selected Papers, Vol. I*). By that time a succession of analysts, starting with Alberto Pedro Calderón and Antoni Zygmund, followed by Joe Kohn, Louis Nirenberg, and Kurt Friedrichs, had developed the very important tool of "pseudo-differential" operators, which is sometimes called "microlocal analysis". Pseudo-differential operators are an extension of differential operators using the Fourier transform. Somewhat as the Fourier transform turns a differential operator with constant coefficients into a multiplication operator, the pseudo-differential operator is approximately the result of transforming a differential operator with variable coefficients into a sort of multiplication. The correspondence is no longer exact, but the error is negligible or controllable. Peter's close friend Louis Nirenberg was one of the main creators of the pseudo-differential operator calculus.

Hörmander abstracted and generalized Peter's work on diffraction to a general theory he called Fourier integral operator calculus. "The purpose of the present paper is not to extend the more or less formal methods used in geometrical optics but to extract from them a precise operator theory...we only use the simplest expansions which occur in geometrical optics" (L. Hörmander, Fourier integral operators, *Acta Mathematica* 127, 1971, p. 81). Hörmander departs from physics to create a far-reaching part of manifold theory and functional analysis.

A clear exposition of this work of Lax is paragraph 5, chapter 6, volume 2 of the English translation of "Courant-Hilbert", *Methods of Mathematical Physics*. My name appears in the preface to this book because I had the honor of serving Richard Courant as a proofreader. I worked on only the first four chapters. The theory of wave propagation (hyperbolic equations) is in chapters five and six. Paragraph 5 of chapter 6, on geometrical optics, was obviously written by Peter Lax.

Becoming Peter Lax's student

I had been admitted to NYU's graduate math program in 1957. In 1958 I signed up for Theory of Functions of a Real Variable taught by Peter Lax. That course, a basic requirement for all math grad students, is a combination of measure theory, point-set topology, and introductory functional analysis. I loved Professor Lax's way of penetrating to the essentials, tossing aside as "garbage" the terms of lower order. He seemed to be talking to the class the same way he talked to colleagues, indifferent to our ignorance and naiveté, as if every student in the room would soon be a real mathematician. I sat in the back of the room and tried to stay inconspicuous, but my homework must have been above average, because one day Professor Lax unexpectedly approached me. Would I be interested in becoming his thesis student? I said, "All right."

I was attempting graduate work in math at age twenty-nine after ten years outside academia. My bachelor's degree was in English lit. I had all my life steered clear of professors, platoon sergeant and drill sergeant, managing editor, shop foreman, or any other "boss". I was born and raised way up in the northeast Bronx. My parents, from Poland and the Ukraine respectively, had never been to high school. Before her marriage, my mother supported herself and her ailing younger sister by operating a sewing machine in a curtain factory. I knew nobody anywhere near Central Park West. I was nervous and naive, not open to friendship with a famous professor a year older than I. In those days, half a century ago, my connection with Peter Lax didn't go much beyond the purely mathematical.

In that year, *Esquire* magazine chose two young American mathematicians as the most brilliant of their generation—John Milnor and Peter Lax. *Esquire* wrote, "Another mathematical prodigy is 32-year-old Peter D. Lax, whose elegant solutions of classical problems in differential equations are admired for their 'light touch'." (I include in this book a picture from that issue of *Esquire*).

It was only fifty years later, while working on this book, that I learned that while I was his student, Peter was helping Chandler Davis, my good friend from undergraduate radical days. Chan and I as students in the 1940s had been involved in active opposition to racism and fascism. After becoming an assistant professor of mathematics at the University of Michigan, Chan refused to serve as an informer to

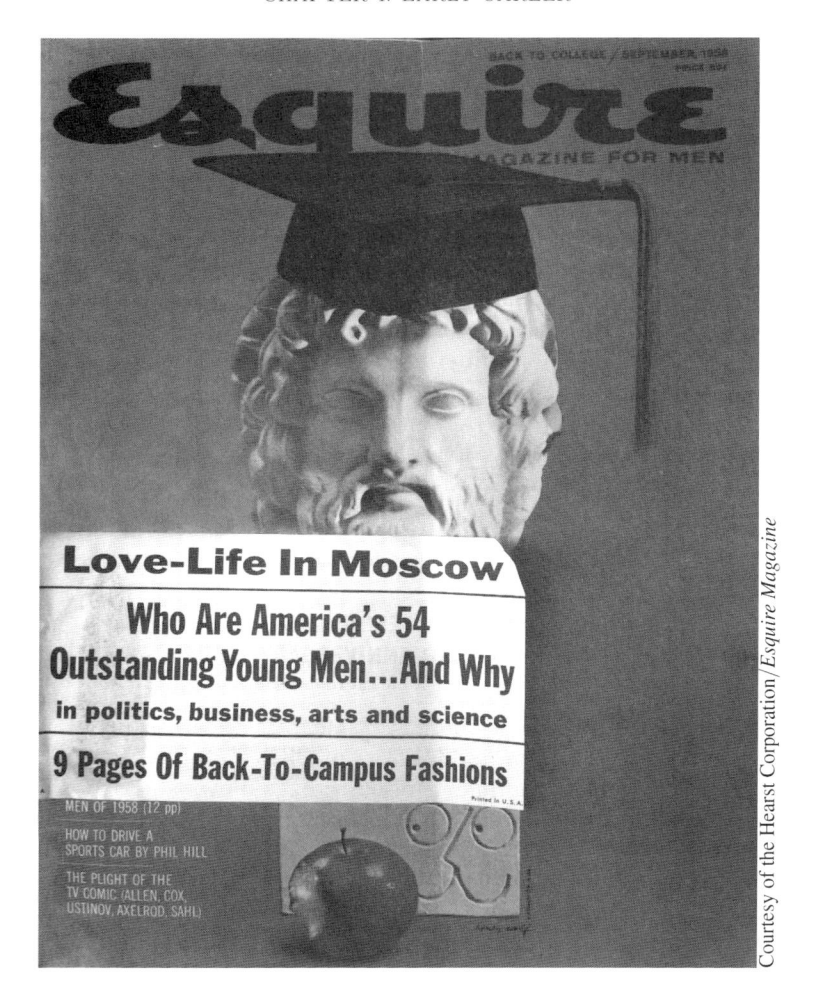

Courtesy of the Hearst Corporation/Esquire Magazine

the U.S. House of Representatives Committee on Un-American Activities. He chose to cite, not the legally advisable grounds of the right to not incriminate oneself, but the more defiant grounds of freedom of speech, proudly mentioning his own ancestor in the American Revolution. Chan served six months in Danbury Prison for contempt of Congress and then was unable to get a teaching job in the United States.

When I started a new path in life by attempting graduate school in math, I went to Chan Davis for advice. He told me to go to NYU, not Columbia. Fifty years later, in the course of writing this book, I first learned that back at that time Chan was in contact with Peter Lax.

Chan writes, "I have great friendship and gratitude for Peter; there are many interests we shared and sometimes discussed. Despite our differences, Peter clearly respected my role as a dissident, and I'm sure that it was at his recommendation that the Courant repeatedly invited me to temporary fellowship. I accepted only once, in spring semester 1962, but the invitations in 1957 and 1958 would have saved my professional life if other things hadn't done so. Peter knows that I appreciated

Peter Lax named by *Esquire Magazine* #11 of America's 54 Outstanding Young Men, September 1958

his help in surviving the blacklist." Peter never mentioned this to me. He didn't know that Chan and I had been friends since 1943.

Even before I decided to try graduate study in mathematics, I had discovered that masterpiece *What Is Mathematics?* by Richard Courant and Herbert Robbins. What a lucky break, to find myself now working as Courant's copy editor! I loved the mathematics itself. I delighted in the open, welcoming environment at the Courant Institute. But I suffered an uncomfortable ambivalence toward applied mathematics. The military connection, with its loyalty oath, security clearance, and atom bomb aspect, was repellent. I never considered applying for security clearance, let alone employment with any military or nuclear weapons employer. It was an anomalous and ambiguous situation for me to produce research that had potential utility in those very places!

Nor did I have the aptitude or connections to find someone else interested in turning my theorems into computer code. Peter Lax and Paul Garabedian, as ranking professors at the Courant Institute, got help turning their algorithms into working programs. I never even knew about such arrangements. I was drawn

UNIVAC Installation at Courant Institute, 1952.
L-R: Peter Lax, Gen. Willoughby (adjutant of Gen. MacArthur), Lazer Bromberg (Director of Courant Computing Center), Gus Kinzel (Chairman of Courant Council), Richard Courant, James Rand (CEO of Remington Rand), Gen. MacArthur, Henry Heald (President of NYU), Gen. Groves (Head of Manhattan Project), Gen. Howley (Commander of Berlin airlift, Vice President of NYU)

by my mathematics to the edge of actual computations, but then shut my eyes and turned my back on such possibilities. This repulsion was virtually automatic, half-subconscious, never even consciously thought out and decided.

One day, after I had been struggling fruitlessly for months, Peter stopped in to my office to say hello. There he saw, taped on the wall behind my desk, a clipping I had cut out of *Life* magazine. From Michelangelo's Sistine Chapel, it was the sinner, aghast as he is condemned to burn eternally in Hell. In heartfelt tones, Peter Lax said, "I feel so sorry for you!" I nodded silently.

After giving up on two earlier problems, I succeeded on my third try. I found all correct boundary conditions to go with any first-order hyperbolic system of PDE's with constant coefficients. My solution even turned out to work much more generally, for arbitrary-order constant-coefficient systems that don't even have to be hyperbolic or parabolic. They need be merely what's called "correct in the sense of Petrovsky". For example, the time-dependent constant-coefficient linear Schrödinger equation is neither hyperbolic nor parabolic, but it is correct in the sense of Petrovsky. I tried to think of a catchy name for systems that are correct but neither hyperbolic nor parabolic, and I came up with "diabolic". But Peter advised against it, so I dropped that idea.

CDC 6600 on loan to Courant Institute from Atomic Energy Commission

Three Wise Men at Stanford: Michael Atiyah, Ralph Phillips and Donald Spencer

With my Ph.D. I moved from New York to Stanford University, with a two-year instructorship thanks to the support of Peter and his collaborator Ralph Phillips, Stanford professor. There I met the king of PDE, Lars Hörmander, and Bob Osserman and Sol Feferman and Paul Cohen, the famous conqueror of the Continuum

Stanley Osher

Hypothesis, with whom I collaborated on a *Scientific American* article. From there to a tenure-track job at New Mexico, where I stayed for nearly half a century, continuing after retirement to teach part-time. When I arrived at the University of New Mexico in 1964, its math department was dominated by statisticians and probabilists, so Peter advised me to move from PDE into probability. That actually happened in a successful collaboration with Richard Griego. We invented a useful new gadget combining linear operators with stochastic processes. It was Peter Lax who suggested the right name for this thing: "random evolution". I was fortunate in collaborating with other probabilists: Mark Pinsky, George Papanicolaou, Bob Cogburn, and Priscilla Greenwood.

In 1958 Peter served as a Fulbright Lecturer in Germany. He lectured on partial differential equations at the University of Göttingen.

That year a student from New Mexico named Burt Wendroff finished his degree at NYU with Peter Lax. Wendroff was one of the two students with whom Peter continued to work after they did their Ph.D.'s. Wendroff had a long career at Los Alamos, specializing in numerical analysis of waves and flows. He met Peter when the lab granted him a leave to study at the Courant Institute. The result was the famous Lax-Wendroff Method and Lax-Wendroff Theorem.

In 1960 he and Peter published the first of their two articles on the "Lax-Wendroff" method for computing nonlinear wave propagation (hyperbolic systems of conservation laws). Their method remains a standard method for these problems. I write about this topic in Chapter 10.

In 1962, the same year I finished at NYU, Peter published two major articles on scattering theory, the topic that was to become a decades-long collaboration with Ralph Phillips. Phillips was famous as the co-author with Einar Hille of the monumental classic *Functional Analysis and Semigroups*. Lax and Phillips's first article on scattering is called "The wave equation in an exterior domain". With Professor Cathleen Morawetz of NYU, they also published that year "Exponential decay of the wave equation in the exterior of a star-shaped object". Scattering theory is discussed in Chapter 10.

In 1963 NYU replaced its UNIVAC by the CDC (Control Data Corporation) 6600. Designed by the famous engineer Seymour Cray, the 6600 outran other computers of its day by a factor of 10. With active involvement of Peter Lax, Richard Courant got one of the first 6600's for his institute when NYU was chosen as a computing center for the U.S. Atomic Energy Commission. The computer belonged to the Atomic Energy Commission.

In 1964 Richtmyer left NYU for Boulder, Colorado, whereupon Peter became director of the computer center.

In the summer of 1965, Peter got me invited to a deluxe summer meeting at the Battelle Institute in Seattle, where I met important people like Jean Leray, Lars Gårding, and Freeman Dyson. He also helped two of my UNM graduate students, Larry Bobisud and Steve Wollman, to receive postdoctoral appointments at NYU.

In 1966 two of Peter's very successful students, Alex Chorin and Stan Osher, became Ph.D.'s.

"Stanley Osher was really my student, although he did his Ph.D. with Jack Schwartz. I got him interested in numerical calculations. He did a very clever thing on the stability of difference schemes and initial boundary value problems. He used difference operators in a very imaginative way. Then he got into scientific computing. He worked on the problem of image reconstruction, refocusing blurred images, that's a very important subject. I think he even has a commercial venture in that."

Osher introduced methods widely used in computational fluid dynamics and related fields, including level-set methods for computing moving fronts, with James Sethian, approximation methods for hyperbolic conservation laws and Hamilton–Jacobi equations, total variation and other PDE-based image processing techniques with Leonid Rudin and Emad Fatemi, and high-resolution numerical schemes to compute flows having shocks and steep gradients with Amiram Harten, Bjorn Engquist, Chi-Wang Shu and others, and Bregman iteration for compressed sensing-related problems with Woiao Yin, Donald Goldfarb, Jerome Darbon and Tom Goldstein.

He has been a thesis advisor for at least fifty Ph.D. students. He was a plenary speaker at the International Congress of Mathematicians in 2010, gave the John von Neumann address at the SIAM annual meeting in 2013, and in 2014 received the Carl Friedrich Gauss Prize for Applications of Mathematics from the International Mathematical Union and the German Mathematical Society for "outstanding mathematical contributions that have found significant applications outside of mathematics." He is a member of the National Academy of Sciences.

In his interview for the SIAM history archive, Phil Colella asked, "What did Osher actually do for his thesis?"

Alexandre Chorin

Photo by Esther Brass-Chorin. Courtesy of Alexandre Chorin.

Peter answered, "It was something in operator theory." (Operator theory is a branch of pure mathematics well known as a specialty of Jack Schwartz.)

"Oh, I see," said Colella. "So you gave him a push in the right direction." Peter answered, "Not push, pull."

Colella's adviser, Alexandre Chorin of the University of California at Berkeley, is one of Peter's most successful students, well known for contributions to computational fluid dynamics.

"Yes," says Peter, "one of my favorite students. I cannot be quite objective about him, he has done wonderful things. One of his early discoveries/inventions was the vortex blob method, which was used by many, including Charles Peskin on his heart models. That was a great idea. For many years he beat his head against the problem of turbulence, but he had very good ideas there also." Chorin received the Norbert Wiener Prize in Applied Mathematics in 2000 "for an outstanding contribution to applied mathematics in the highest and broadest sense."

I asked Chorin to comment, and he wrote, "I owe Peter a lot, more than I know how to put down on paper. I approached Peter and asked him to be my thesis adviser in May or June 1965. I had been having a lot of personal difficulties and health problems and had found it difficult to turn up for classes and study regularly; I got some B's and even a C in my classes. I had been working with another mentor who saw I was in difficulty and tried to help me by supervising me more closely. But this didn't work. I had been told officially that I wasn't making

Jeffrey Rauch

good progress towards a Ph.D. and that my fellowship, and even my visa, were at risk. I had had a sort of relationship with Peter, having taken his real variables course and done poorly and having three conversations with him or, rather, one conversation three times. Each time he asked whether I was Hungarian (Chorin being the name of a distinguished Hungarian family), and I explained, no, I was born Choroszczanski. "Chorin" is a hebraization of the first four letters of my previous name that means "freedom" in Hebrew.

"Peter reassured me about my fellowship. I knew some fluid mechanics from my days as an engineering student; I had a job where I learned to program (which was not that common at the time), and I had taken a numerical analysis course at NYU. I have some recollection of hearing a seminar lecture on thermal convection and feeling that I could do a better computation. Peter suggested papers to read. It is very clear to me that Peter had some idea that I was having difficulties not necessarily related to mathematics and was very patient with my absences and my not always turning up when summoned. I was grateful then for Peter's patience and humanity, and am even more grateful to him now that I have had grad students of my own. Without his kindness I doubt I would have received a Ph.D. It is not easy to see the need for kindness when the usual view of one's duty is to insist on steady progress.

"I looked at thermal convection and came up with a solution method ('artificial compressibility'). Peter asked whether I was interested in staying at Courant, and I was extremely happy to do so. My years as Peter's postdoc were among the happiest of my life. I worked a lot, and Peter was extremely supportive and full of suggestions. But after my son was born, I became impatient with living in New York and wanted to move. Peter's help was crucial: he didn't think I should move, and when I made a poor choice of a place to move to, he told me not to do it. He introduced me to Mark Kac, who helped arrange a one-year visiting appointment in Berkeley, where I have stayed."

When Peter learned of Chorin's account of their interaction, he had a simpler explanation. Alex previously had been working under Joe Keller, who likes to manage his thesis students closely. "Do this, then do that." Alex was unhappy with close supervision. When he came to Peter, Peter simply let him do whatever he liked. Great success!

Another successful student of Peter's is Jeffrey Rauch, who received his Ph.D. in 1971. He has long been a professor at the University of Michigan, where he continues to work in partial differential equations and wave propagation.

His website reads, "My recent work has included the short wavelength asymptotic analysis of solutions of linear and nonlinear hyperbolic problems. Also, the analysis of nonreflective and/or absorbing layers in numerical methods. And precise finite speed of propagation estimates for linear problems. The quasilinear versions are outstanding open problems, and so is the mathematics of high powered lasers."

In response to my request for his reminiscences of Peter Lax, Rauch writes: "I'd been told as an undergraduate that Peter Lax would be a wise choice for advisor. I took a hyperbolic PDE course from him that closely resembled his book now published. The treatment of progressing waves and propagation of singularities particularly appealed to me, along with the singular integral/pseudodifferential work. He suggested that I look at mixed problems using methods he'd developed with Friedrichs. Kreiss visited and I started talking with him about the problem that would become the heart of my thesis. Peter kept me on a long leash. He had four students about my age: Keyfitz, Goodhue, Homer Walker, and me. It was interesting to see how he handled us differently. Those who needed more tending got it. I learned a ton from him in informal conversation: "do you know about..." or "have you seen...". There are things he told me that I pass on to others nearly verbatim.

"After the first week of November, he told me that I should write up a thesis to defend in December! I was a bit stunned, but it had the good effect of making me write without any frills. No summary of known results or anything like that. Theorem statements, proofs and examples, absolutely no frills. Forty pages. It was good advice. Once the thesis was written, I could move on. I proved the best stuff in the six months after the defense.

"When I proved my first little theorem I was extremely excited and wanted to tell someone, anyone. But there was a joint NYU-Japan meeting going on in Japan at the time and EVERYONE was gone. Joel Smoller was a visitor from Michigan and kindly listened, and I've been at Michigan ever since.

"When Peter came back he was very busy, and I couldn't arrange a meeting with him. After a week of rebuffs, I wrote him asserting that I was beginning to doubt his existence. Then he did make an appointment with me, but when I was ushered into his office, he wasn't there! He soon emerged from hiding under his desk.

"Of course he fell asleep several times in embarrassing situations. Once I went with him to see a film. He serenely slept through the whole thing... except that he laughed at the jokes!

"This fits with everyone's experience of him sleeping through seminars, only to ask good questions at the end.

"I loved the way he would coax facts from Louis Nirenberg. Louis would insist that he didn't know the answer, and then, after appropriate prodding from Peter, he would produce the required information.

"At my thesis defense, he brought a bottle of champagne and had the bottle signed by the committee members. I still treasure it.

"I never got to play tennis with him, never sailed with him, never visited Loon Lake. As I write these quick notes I am saddened by Anneli's death."

In 1966 Peter received the first of two Lester R. Ford awards from the Mathematical Association of America for an outstanding expository article, "Numerical solutions of partial differential equations" (*The American Mathematical Monthly*, vol. 72 (1965), Part II, pp. 78–84).

He also published in that year another influential, much-quoted article jointly with his close friend Louis Nirenberg, "On stability of difference schemes: A sharp form of Gårding's inequality".

One morning early in my years as Peter's thesis student, I entered his office to find him glowing in smiles. "Louis is back!" he cried out. Louis, I wondered? Oh yes, Louis Nirenberg, also one of the PDE specialists at Courant. He had been on leave in Italy; now he was back home! Louis Nirenberg and Peter Lax were graduate students together at NYU. Then they both stayed on to become famous faculty members there—Louis, a world master at elliptic partial differential equations, and Peter, a world master of hyperbolic PDEs. They hardly ever collaborated or produced joint publications. But their conversations and their intellectual and emotional interactions were a vital part of their creativity and success (*Loving + Hating Mathematics*, p. 138).

I interpolate here a brief account of Louis Nirenberg's life and mathematical work. Nirenberg was born in Hamilton, Ontario, where his father was a teacher of Hebrew. The home language was Yiddish. Louis's Hebrew tutor enjoyed math puzzles, and the two of them spent half their time together secretly doing puzzles. He grew up in Montreal and regrets that he never had a French-speaking friend. The Jews there were a tight-knit subset of the English-speakers. He has lived on the Upper West Side of New York since 1949, and he loves his apartment, his neighborhood, and his city. He also loves to travel. He is particularly in love with Italy. He was a happy participant in the first big Soviet-American joint mathematics conference in 1963 in Novosibirsk. He says it was like being on an ocean voyage where you get to know many new friends intimately.

Interesting problems in partial differential equations rarely are solved by an explicit formula. The essential point usually is showing that a proposed approximation procedure actually converges to a solution. Machinery from functional analysis proves convergence if appropriate estimates control the approximation.

For some elliptic equations, the classic maximum principle provides the needed estimate. Nirenberg is quoted as saying, whether in jest or in earnest, "I have made a living off the maximum principle." But other kinds of estimates may be required. The functions in question, along with their derivatives, may be squared or taken to other powers and then integrated against appropriate "weight functions". Proving the strongest estimate with the appropriate norm is the technical essence of the modern theory of partial differential equations. Mathematicians came from everywhere to seek Nirenberg's counsel on questions about estimates. They sometimes made errors in their calculations, but Nirenberg had a nose for their errors! Now in his eighties, he makes errors himself and has to correct them before publication.

Nirenberg and Joe Kohn created the theory of pseudo-differential operators, an extended version of the Calderón-Zygmund singular integral operators. This tool enabled mathematicians to handle differential operators with variable coefficients in a way analogous to the traditional Fourier transform methods for operators with constant coefficients. We discussed these operators previously in relation to Peter's famous paper on diffraction, and will return to them in Chapter 8, in Peter's book on hyperbolic equations.

Louis contributed these memorable incidents involving Peter's unfortunate problem of narcolepsy:

"Once we stayed up on election night playing bridge, he and I and Anneli and my wife. He was the best bridge player in the group, but sometimes after he played a card he fell asleep before playing the next.

"Once when Olga Oleinik gave some talks at the institute, she complained that Peter fell asleep during her lectures. At her next lecture Peter wrote out something and circulated it. It said, "We the undersigned testify that Peter has slept through some of our best lectures." Everyone (about a dozen people) signed it, and Peter gave it to Oleinik at the end of her talk. I suggested she frame it, but I doubt that she did.

"I met a student of his in Taiwan who told me that he went to Peter's office to get his signature on his Ph.D. dissertation. But Peter objected, saying, 'Why did you get it typed up before we went over it together to make sure it was OK?' So the student went to the board and started lecturing about his research. Peter quickly fell asleep. The student waited quietly till Peter woke up and then handed Peter the cover page and a pen. Peter signed it.

"Peter once said that at the time when a certain seminar was usually scheduled, even though the seminar had been canceled, Peter felt sleepy for that hour."

Louis has great admiration for Peter's work. "It is elegant and deep, shows wonderful taste, especially his work with Ralph Phillips. He writes beautifully." Their one joint paper was done by mail while Louis was in Japan.

Nirenberg concluded our interview by saying, "I must say all the people I've worked with have been extremely nice. It's one of the joys of working with colleagues. Peter Lax seems like a brother to me. That's the thing I try to get across to people who don't know anything about mathematics. What fun it is! One of

James Glimm

the wonders of mathematics is you go somewhere in the world and you meet other mathematicians and it's like one big family. This large family is a wonderful joy."

Reading these comments of Nirenberg's, Peter pointed out that during the Cold War, when official suspicion and hostility made any contact between U.S. and Soviet citizens problematic, many mathematicians in the two countries became close friends and collaborators. Two Russian examples were the leading contributors to partial differential equations, Olga Oleinik and Olga Ladyzhenskaya.

Nirenberg's mathematical family includes 46 Ph.D. students and 331 mathematical descendents. Peter has been like a brother to Louis, and Peter's sons considered Louis an uncle.

In 1967 Peter published with his NYU colleague James Glimm an announcement of major new results, "Decay of solutions of systems of hyperbolic conservation laws", introducing *random* choice of data as a new device in this field of analysis. That year he also published a major research treatise, *Scattering Theory*, with Ralph Phillips. It would be reprinted in 1989.

In 1968 he published a very influential article clarifying the remarkable phenomenon of "solitons", wave forms that persist in nonlinear interactions. (This topic also is elaborated on in Chapter 10.) That year he delivered the von Neumann Lecture to the Society for Industrial and Applied Mathematics.

In 1970 he published with James Glimm *Memoir 101 of the American Mathematical Society*, "Decay of solutions of systems of nonlinear hyperbolic conservation laws". It contains the full calculations and proofs of the results announced in 1967. "It was a hundred pages and it is very painful to read. It was even more painful to write. Fortunately, there's a simplified version in the dissertation of Konstantina Trivisa, a student of Constantine Dafermos. Glimm's idea is a landmark in this field. It provided estimates that led to renormalization groups. That's what's missing in the theory for two dimensions. It would take another Glimm to come up with such a clever idea. It's a scandal that nothing is known with rigor for flows with shocks in two dimensions. We have a great deal of numerical experience, and this is all to the good. There are test problems like the Riemann problem or flow with a partially obstructed channel that have been calculated repeatedly by three, four, five quite different methods, and the resulting flows are remarkably in agreement. They must be the truth. The truth is known, so why can't we prove it?"

CHAPTER 5

The famous CDC 6600 bomb-scare adventure

In May of 1970, NYU's CDC 6600 was kidnapped.

In 1968 Richard Nixon had been elected president and promised to end the Vietnam War. In November 1969 it was revealed that at the Vietnamese village of My Lai, American troops had massacred an estimated 504 women and children. The following month the first draft lottery since World War II was carried out. Campuses across the country erupted in protest.

Then on Thursday, April 30, 1970, President Nixon announced the "incursion" into Cambodia by U.S. combat forces. Shock and outrage flared up on many college campuses. The next day, Friday, May 1, at Kent State University about 500 students demonstrated on the Commons, a grassy knoll in the center of campus, against Nixon's expansion of the war. Students watched while a graduate student burned a copy of the U.S. Constitution and another student burned his draft card.

At NYU, President James M. Hester called an emergency meeting of the University Senate on Sunday evening, May 3. An overwhelming majority condemned

"They Can't Kill Us All" —NYU student response to massacre of four students at Kent State University, May 4, 1970

Jürgen Moser (1928-1999)

the invasion of Cambodia and the renewed bombing of North Vietnam and called for speedy withdrawal of American forces from Southeast Asia. They urged all departments of the university to hold meetings on Monday and Tuesday from 12 noon to 2 p.m., to develop concrete proposals for bringing effective pressure on the United States government to end the war in Southeast Asia and called for a mass meeting of the New York University community on Wednesday, May 6.

But the possibility that the NYU administration and faculty could lead the university in an antiwar effort was destroyed on the next day, Monday, May 4, in Kent, Ohio. At an antiwar demonstration on the Kent State University campus, National Guard troops fired into a crowd of students hundreds of feet away, killing four students and wounding nine. Two of the students killed had been participating in the protest. The other two had been walking from one class to the next at the time of their deaths.

These shootings of students brought four million students into protest nationwide. Student strikes closed over nine hundred American colleges and universities. Kent State University remained closed for six weeks. At New York University a

Martin Davis

Photograph by Virginia Davis

banner hung out of a window, reading "They Can't Kill Us All." On May 8 eleven student protesters at the University of New Mexico were bayoneted by the New Mexico National Guard. On November 9, 100,000 people demonstrated in Washington, D.C., against the war and the killing of unarmed student protesters. Ray Price, Nixon's chief speechwriter, recalled that day by saying, "The city was an armed camp. The mobs were smashing windows, slashing tires, dragging parked cars into intersections, even throwing bedsprings off overpasses into the traffic down below." President Nixon was taken to Camp David for two days for his own protection.

Monday evening, May 4, at a meeting of around 1,500 students, faculty, and others in NYU's Loeb Student Center, many speakers denounced Nixon's Cambodia war and called for "taking over" New York University. In the following days the administration and faculty simultaneously defended the university from being "taken over" while actively opposing the war.

It was the first week of May, classes were almost over. Many professors and administrators shared the students' anger over the "incursion" into Cambodia and

Mass protest at Warren Weaver Hall, May 7, 1970

the killings at Kent State. On Tuesday, May 5, President Hester suspended classes "to enable students and faculty members to engage in joint efforts to influence national policy." He said that "the Senate concluded that this is far more effective in pursuing the cause of peace than staying home." He invited proposals for constructive action to affect national policy. Most of NYU's colleges canceled final exams. Students received Pass or Fail instead of letter grades.

Classes had been suspended in order to allow students to work actively against the war, but most of the students just took the opportunity for an early start on summer vacation. The general population, outside the universities, did not join in the protests. Some even actively opposed them.

At NYU three buildings were seized by activists: the auditorium of the Loeb Student Center, which became the site for "mass meetings"; Kimble Hall, where the university had printing facilities that were taken over by the "strikers"; and Warren Weaver Hall, home of the Courant Institute and its CDC 6600 computer.

At 11 o'clock on Tuesday morning, May 5, about one hundred strikers walked into Warren Weaver Hall and tried to get into the computer center on the second floor. When they couldn't get in, they threatened to "take other measures" unless the computer was immediately shut down. The second floor was then cleared

Molotov Cocktail used in attempt to blow up CDC 6600 computer at Courant Institute, May 1970

and secured, and members of the Courant faculty engaged in discussions with the strikers on the first floor of the building. After an hour, about sixty of the strikers tried again to get into the computer center.

The director of the Courant Institute was Jürgen Moser, a world leader in the study of dynamical systems and also, years later, a brother-in-law to Peter. Moser tried to defend the locked door between the staircase and the second floor. "I attempted to prevent this group from breaking the door at the top of the stairway. For my efforts, I was pushed and shoved around and was unable to deter them. The strikers proceeded to break the door and gain access to the hallway outside the anteroom to the computer room. Once in the hallway, they broke the glass door to the anteroom, which contains keypunch machines and other machinery used in computer research projects. Fortunately, the door leading from the anteroom to the computer room itself is made of steel and was locked, thereby barring entrance to this group of people.... During the afternoon and early evening of May 5, this group proceeded to deface the walls and ceilings of the stairways, halls, and rooms throughout the entire building. Paint was spilled on rugs and parquet floors on the thirteenth floor, windows were broken, and doorknobs removed to gain access to certain rooms."

Professor Martin Davis writes, "I remember those events very well. A novelty at NYU was the demand that NYU put up $100,000 dollars bail for an imprisoned member of the Black Panthers. I was present at a discussion of the proposal to occupy Warren Weaver Hall. I was allowed to speak against it, but the wind was blowing against me. I rushed to my office on the ninth floor, and left clutching documents I feared might be lost. As I left the building, students were rushing in."

At 5 o'clock that afternoon, Chancellor Alan M. Cartter told the Faculty Senate that there had been "a minor confrontation" between the strikers and Courant staff and faculty. He said the two groups had held lengthy discussions and the Courant faculty agreed with some of the strikers' objectives. The university administration was hoping the strikers would be persuaded to leave the building or at least to do no more damage and keep it open to faculty and staff.

But the strikers managed somehow to get into the computer room. On the next afternoon, Wednesday, the university administration received this telegram:

> We, as members of the N.Y.U. community occupying the Courant Institute, are holding as ransom the Atomic Energy Commission's CDC 6600 computer. At a general meeting in Loeb Student Center, the people put forth the following demands: the University must pay 100 Thousand Dollars to the Black Panther Defense Committee for bail for one Panther presently held as political prisoner in New York City. Failure to meet this demand by 11 a.m. Thursday, May 7, will force the people to take appropriate action. In addition, if the University Administration should call in police or other authorities, the above action will be taken immediately. In the meantime, no private property will be destroyed.
>
> [Signed] *N.Y.U. Community on Strike*

Martin Davis and Mel Hausner wrote a leaflet urging the strikers to spare the computer, drop the demand for bail money, and focus on ending the war. But Martin wasn't allowed back into the building. His wife, Virginia, was allowed to enter with the leaflets and found "first floor people" and "second floor people". On the first floor, the rank and file. On the second floor, "a highly disciplined group strictly controlling access to the computer room."

Peter wrote, "Several of us felt that the police should have been called immediately. But all this happened a few days after the shooting at Kent State, so the NYU Administration's reluctance is understandable."

On Wednesday evening, in Loeb Student Center, about one thousand people were told that "Warren Weaver Hall has been captured and liberated." The meeting passed a motion demanding $100,000 ransom for the computer, with an amendment that if the ransom was not paid, the university would not be returned to its president.

The report of the NYU News Bureau goes on, "Then Dr. Robert Wolfe, an assistant professor of history, took the microphone and said, "You have voted with your voices; now vote with your feet! On to Courant!"

A portion of the audience left the auditorium, proceeding to Warren Weaver Hall. Jürgen Moser was there in the computer anteroom. A group of strikers ordered him to leave and not return. "When I objected and refused to move, I was grabbed by four male members of this group who forcefully attempted to evict me. George Morikawa, a research scientist at the institute, prevailed upon me not to resist."

A nine-member "liaison committee" of Senate members was appointed by President Hester and met with strikers at around 6:30 p.m. on Wednesday. The senators pointed out to the strikers that they were committing extortion. The strikers insisted that the university work out some way of raising $100,000. "After about an hour's fruitless discussion, the meeting broke up."

Then a group of senators offered to raise a bail fund for the Black Panthers, which could be used for scholarships for black students after it had served bail purposes. The strikers pressed them to say how much money they would raise by 11 a.m. Thursday, but no estimate could be given. The Senate liaison committee and the strikers resumed their meeting at 11 p.m. Wednesday, and went on all night, ending fruitlessly at 5 a.m. Thursday.

President Hester then issued a statement: "The computer in the Courant Institute contains the professional work of a number of distinguished scholars and is the property of the Federal Government. Years of effort would be destroyed by damaging this machine. We have no choice but to protect the work of these scientists and the computer. We must exercise our legal and moral responsibilities to evict those persons who are illegally occupying the Courant Institute."

Two final appeals to leave Warren Weaver Hall peacefully were made by Dr. Roscoe C. Brown, director of the Institute of Afro-American Affairs, offering pledges of from $10,000 to $15,000 toward the Panther bail fund. The strikers told Dr. Brown, "Too bad. This is just not enough. There is nothing we can do."

An injunction signed by Justice Joseph A. Sarafite of the Supreme Court of the State of New York was delivered to the occupiers. Singing high school kids massed in front of the building in disciplined groups. Professor Davis turned to Professor Hausner and said, "I see now that I have the soul of a Menshevik." (In Russia in 1917, the "Menshevik" faction of Socialists opposed the Bolsheviks' forcible seizure of power.)

Between 10:30 and 11 a.m., Chancellor Cartter received a message saying that the strikers had decided to leave and hold a press conference without harming the computer.

But according to Bruce Kayton (*Radical Walking Tours of New York City*, Seven Stories Press, 2003, page 29), there had been a meeting at which an anarchist group, the "Transcendental Students", proposed erasing the computer tapes with magnets, while the Marxist Students for a Democratic Society (SDS) wanted to blow up the computer with Molotov cocktails.

At noon, Martin Davis writes, the occupiers marched out in a disciplined manner, joining around a thousand people in front of the building.

Chancellor Cartter and Peter Lax, together with assistant professors of math, Emile Chi and Frederick Greenleaf, rushed into the building. Peter said, "I was in the lobby with a number of Courant colleagues from the Computing Lab—I was director of the Lab—and when the occupiers left, I could smell smoke and said, 'Let's run up and see what's going on.'"

Smoke was coming under the door to the staircase, but the doorknob had been removed. When the door finally swung open, the three mathematicians rushed upstairs to the computing center.

Greenleaf and Chi are still in New York, forty-three years later, at NYU and at Staten Island University respectively. So far as I know, this is the first time their account has been published.

Professor Greenleaf writes, "The doors to the area containing the CDC 6600 had been disabled and could not be opened. There were two small windows, about 7×6 inches. No one could squeeze through them, and they were set too high for anyone to reach the inside door knobs (which might not have worked in any case). Some floor tiles had been removed, creating a narrow channel under the disabled doors. A really dorky fuse fashioned out of a long stretch of twisted toilet paper had been burning for a minute or two before Chi and I arrived. It had burned its way through the channel under the door. Through the door windows we could see it SLOWLY making its way toward large jars of presumably inflammable material on top of the computer. Chi and I ran to the stairwell and grabbed an old-fashioned fire extinguisher (the old acid/soda type filled with water). Back at the door, one of us spotted through the window, while the other pushed the extinguisher hose under the door and aimed a stream of water according to instructions from the spotter. (I can't remember which of us did the spotting and who held the hose.) We were close enough to douse the fuse, and that was the end of the danger."

Chi adds: "Fred was very slim and I was somewhat overweight. Fred squeezed part way through the cable channel under the door and held the hose, while I directed his aim by looking through the window. I remain somewhat amazed that both of us had the nerve to do this instead of running out of the building! Moments later Peter Lax and Jürgen Moser and a bunch of cops came charging up the stairs to survey the damage, but by then the real excitement was over." Peter says, "The rest of us rushed into the computer room and removed bottles tied to the computer, filled with flammable liquid. I have a photo of the bottles."

Years later, in a blog for Stuyvesant High math alumni, Dan Kotlow (class of 1961), reminisced:

"Did you know that in the early '70s some malcontents tried to blow up the NYU computer center, and Peter Lax stamped out the fuse?" [This, of course, is incorrect.] Kotlow goes on, "Professor Chi, by the way, was the Stuyvesant valedictorian in 1953, and lent us a document prepared by NYU several months after the incident.

The NYU News Bureau report goes on, "The computer room is located directly over the main entrance way of Warren Weaver Hall. According to bomb squad experts, a 50-foot fuse to the bombs had been intended to burn for about 1 1/2 minutes before igniting the explosives—time to allow the occupiers to exit and reach safety from the explosion. But the gasoline had partly evaporated, so it took over two minutes for the flame to reach the door to the computer, where it was extinguished. Thus a matter of seconds stood between the discovery of the burning fuse and serious injury to more than 1,000 persons, both outside the largely glass building and in the hallway a few feet below.

"It was later determined by bomb squad experts that the bombs would almost certainly have gone off. The explosion could have shattered glass windows, showered broken glass and flaming gasoline on the crowd outside, and caused heavy damage to the computer."

Peter wrote, "When I reconnected with Anneli afterwards she had this to say: 'You smell smoke and rush upstairs with a group of people to see what is going on. Are you crazy?'"

"The only thing I could say was, I was so angry that I didn't think."

"Somewhat later," Martin Davis writes, "I organized a Courant 'lobby' to go to Washington to argue against the war. It was mostly students, but I was surprised and pleased that Peter came along. We spoke to both N.Y. senators, some congressmen and others. Of course, we had little, if any, effect."

It remained for the NYU faculty and administration to plan a commencement ceremony in June. On May 14 the NYU Senate passed a motion by Dr. Roscoe

Brown that the commencement ceremony have the theme "Convocation for Peace". A week later, Brown reported that a group of students were demanding that all planning be done by them. Dr. Cartter listed some demands that he had received: "No academic procession. No academic costumes. No alumni citation. No honorary degrees. No speakers except ones they choose. No music except their selection of rock bands and folk singers." The Senate considered whether to hold a commencement at all and decided to go ahead and risk it, using their own committee to do the planning.

On June 9 the Commencement Convocation for Peace took place in Madison Square Garden without major incident, attended by 10,000 students, faculty members, and guests. The program was created by a committee of half students and half faculty and administration. They heard a speech by Nobel Prize winner George Wald, the Higgins Professor of Biology at Harvard (NYU, 1927), very critical of the war in Vietnam and Cambodia.

On July 30, Robert Wolfe, assistant professor of history, and Nicholas Unger, graduate teaching assistant of physics, were arrested and charged with "conspiracy, attempted grand larceny by extortion, and attempted coercion." They pled not guilty. Ten months later on May 27, 1971, they pled guilty to a reduced charge of "attempted grand larceny in the third degree," and were jailed for ninety days.

The U.S. war against Vietnam would continue for three more years, until 1973.

In 1970, after the great computer kidnap adventure, the NYU Senate created a "Committee on University Purposes and Guidelines", to learn from that experience and to help NYU be better prepared for any such future emergencies. Peter accepted appointment as a member of that committee, which became known as "the Hindle Committee", after its chairman, Brooke Hindle of the history department. The University Archive in Bobst University Library contains minutes of eight meetings of that committee between June and November and its report to the Senate on December 1, 1970, which was adopted on February 11, 1971.

The committee did not undertake to "review or evaluate actions taken following the events of May 4, 1970," but rather set out to "single out those ideals which we cherish today and which we expect to increase in value tomorrow...to define the inner defense perimeter which cannot be compromised...and also outline how the university can contribute to life within our own institution and throughout the world...by the solution of problems of today and tomorrow...and contribute to the quality of life."

The Archive contains two memoranda submitted to the committee by Peter Lax on August 13 and September 12. The main thrust of Peter's memoranda is that the university should stay out of politics. Taking stands on political controversies not specifically related to its educational and research missions can only damage the university, which has no special standing or authority on other issues. Getting involved in such issues would likely create divisions within the university and damage its reputation and credibility with the public.

While that position is plausible as a general proposition, it seems that in the fall of 1970 it had specific application to actions in the previous spring by President Hester and the NYU Senate. They had joined in the protest against the Cambodian "incursion" and the Vietnam War. Some alumni and a few faculty were unhappy with the public antiwar position of the university president and senate. Lax's memoranda made no reference to the events of May (which specifically were not part of the assignment of the Hindle Committee), but it seems to me now that Peter's memoranda at that time unavoidably suggest his disagreement, if not condemnation, of the actions of the president and the senate. Yet Hester was merely joining in with forty-eight other presidents of U.S. universities! After all, students had been shot dead on campus by National Guard troops. Could one maintain that the shooting was a proper concern of NYU but the war itself was not? There is nothing to indicate that NYU lost any prestige or image by Hester's actions. Rather the contrary, it might have suffered such a loss if he had refused to join in.

For my part, being in New Mexico at the time, not New York, I was more concerned with the events at the University of New Mexico. On May 5, 1970, a protest against the Vietnam-Cambodian War and the Kent State massacre occupied our Student Union Building. The National Guard were ordered to clear the campus, using fixed bayonets. They stabbed eleven people. Our university president at the time seemed entirely passive. I attended a meeting of the whole university faculty to protest his failure to try to protect the university from the National Guard. Two years later I joined an antiwar march to our local air force base, where we received a dose of tear gas.

Clearly, back in 1970 I would have been in serious disagreement with Peter Lax if we had had the chance to talk about these matters. By now, it's ancient history. I understand how a dedicated, devoted professor can be outraged if his work, his department, his university are disrupted by intruding lawbreaking protestors. I understand his outrage, but I cannot agree with it, not if the matter under protest is critically, immediately, and urgently important for the nation and indeed the world.

CHAPTER 6

Later career

In 1972 Peter delivered the Hedrick Lectures to the Mathematical Association of America.

Also in 1972 he was elevated from director of the computer center to director of the Courant Institute of Mathematical Sciences. He held that post until 1980.

By 1972 the CDC 6600 had become obsolete. As soon as Cray finished the 6600, he started making a faster one to replace it. In 1968 he completed the CDC 7600, again the fastest computer in the world. The 7600 had about three and a half times the clock speed of the 6600 and ran significantly faster than that, thanks to other technical innovations. CDC sold about fifty of the 7600s. Cray left CDC in 1972 to form his own company.

Peter said, "Perhaps my finest contribution was when it became clear that the CDC 6600 was no longer competitive. It had been a supercomputer in its time, from 1964 through the early 1970s. A lot of outside users came to use it on the premises, and they paid for the usage. In-house users would not have justified the expense. The outside users were the basis of the computing center budget. But then other places acquired computers, and suddenly the 6600 became a white elephant. The death blow was the establishment of the thermonuclear fusion network with its own dedicated computer. We lost a lot of customers that way. Max Goldstein and I had the good idea to turn the CDC 6600 over to New York University to become the university computing center, where it was very useful.

"I started my directorship at the worst possible time for New York University. They had just transferred their School of Engineering to the Brooklyn Polytechnic Institute, which recently became a branch of NYU." The tenured members of the Engineering College math department had a legal claim on NYU. Those without tenure were at liberty to seek employment elsewhere. For the time being, they all moved downtown to Washington Square College or to the Courant Institute.

Making room for them at the Courant Institute was a delicate task for Peter. One welcome addition was Joe Keller, the famous applied mathematician Peter had known since their student days, who had been the math chairman at University Heights. But adding a large group of faculty from the Heights restricted him, as director of the graduate school of math, from hiring new faculty. Despite the constraint in hiring caused by the shutdown of NYU's engineering school, Peter was able to make some important hires, which he feels may have been his biggest accomplishments as director of the institute. "I was instrumental in hiring among others Charlie Peskin at the recommendation of Alexandre Chorin. And Sylvain Cappell at the recommendation of Joe Kohn. Both were enormous successes."

Peskin had just earned his Ph.D. at Berkeley under Chorin. He became a leading researcher in the fluid dynamics of blood flow. His "Immersed Boundary Method" handles the coupling between deformable immersed structures and fluid

Joseph B. Keller

flows: in particular, between the heart muscle and the blood flowing past it. That method also successfully modeled the generation of lift in insect flight and wave propagation in the inner ear. He won a MacArthur Fellowship, membership in the National Academy of Sciences, the George David Birkhoff Prize in Applied Mathematics, and a Josiah Willard Gibbs Lectureship at the American Mathematical Society.

Cappell had been a promising tenure-track professor at Princeton but was denied tenure there when Princeton decided they had enough topologists. The Princeton analyst Joe Kohn, a close friend of Louis Nirenberg, urged the Courant Institute to hire Cappell.

In 1972, under Peter's leadership, the Courant Institute created a new computer science department, alongside of the existing mathematics department. Jack Schwartz became the founding chairman of the new department. He and Martin Davis took joint appointments in the two departments. This new computer science department inside the Courant Institute was put at risk by a group of NYU engineers, who proposed starting activity in what they called "informatics". Peter fought very hard to stop that. "I think it would have been very bad for the university to have two computing departments—it certainly would have been very bad for our computer science department."

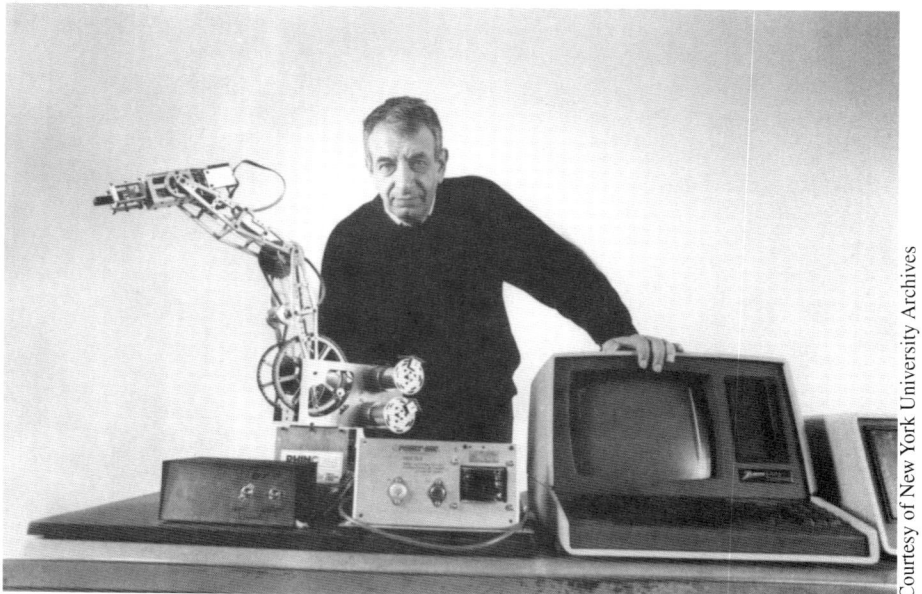

Jack Schwartz (1930-2009)

In 1973 he published a book, *Hyperbolic Systems of Conservation Laws and the Mathematical Theory of Shock Waves*. These two subjects are among Peter's main contributions to both pure and applied mathematics. I discuss them in Chapter 10.

The Mathematical Association of America annually awards two different prizes for expository articles. The Chauvenet Prize, $1,000 and a certificate, has been awarded since 1925 for "an outstanding expository article on a mathematical topic." There's no restriction about the journal in which the article appears. The prize is named after William Chauvenet, a math professor at the U.S. Naval Academy. The Lester R. Ford Awards, established in 1964, are for "outstanding expository papers in *The American Mathematical Monthly*." Ford was a distinguished mathematician, editor of the *Math Monthly* from 1942 to 1946, and president of the Mathematical Association of America in 1947–48. Peter Lax's article, "The formation and decay of shock waves" (*The American Mathematical Monthly*, vol. 79 (1972), pp. 227-241) won both prizes: the Lester Ford Award in 1973 and the Chauvenet Prize in 1974.

This paper describes the origin of the governing equations for the propagation of shock waves, some of their striking phenomena, and a few of the mathematical tools used to analyze them.

In 1974 Peter's favorite student, Amiram Harten of Israel, earned his Ph.D. Tragically, Harten would die prematurely of a heart attack in 1994 at age forty-eight. "His death very much affected me. He was a heavy smoker and he drank twenty cups of coffee a day, and he was overweight except when he went on a dieting binge and so...well, it was a great pity. He was just starting an interesting line of research on multi-resolution. He had other very good ideas, like switching from a second order scheme to a diffusive scheme near a discontinuity—that was his

Amiram Harten (1947-1994) in Palo Alto, 1982

master's dissertation with Gideon Zwas. And essentially non-oscillatory schemes was his idea."

Harten made fundamental contributions to high-resolution schemes for solving hyperbolic partial differential equations. He developed the *total variation diminishing* scheme, which gives an oscillation-free solution for flow with shocks. In 1987, along with Björn Engquist, Stanley Osher, and Sukumar R. Chakravarthy, Harten published one of the most cited papers in the field of scientific computing, "Uniformly high order accurate essentially non-oscillatory schemes, III" in the *Journal of Computational Physics*. It was republished in 1997 in the same journal. In 1990 Harten gave a talk on "Recent developments in shock-capturing schemes" at the International Congress of Mathematicians in Kyoto.

In 1975 Peter Lax won the Norbert Wiener Prize, which is jointly given by the American Mathematical Society and the Society for Industrial and Applied Mathematics.

In 1976 he published two books. One, with Ralph Phillips, was the research monograph *Scattering Theory for Automorphic Functions*, published jointly by Princeton University Press and the University of Tokyo Press. The other was the undergraduate textbook, *Calculus with Applications and Computing*, written jointly with Sam Burstein and Anneli Lax. As an original and useful textbook, it is a success. Commercially, it was a failure.

Peter Lax says, "A thorough revision with coauthor Maria Terrell has appeared in the fall of 2013. We hope and expect that it will do much better than its predecessor."

Reform of calculus teaching has long been a major cause of Peter's. He wrote, in "The flowering of applied mathematics in the United States", *A Century of Mathematics in America, Part II*, American Mathematical Society, 1989, p. 463: "The applied point of view is essential for the much needed reform of the undergraduate curriculum, especially its sorest spot, calculus. The teaching of calculus has been in the doldrums ever since research mathematicians have given up responsibility for undergraduate courses. There were some notable exceptions...but calculus, in spite of some good efforts that did not catch on, has remained a wasteland. Consequently the standard calculus course today bears no resemblance to the way mathematicians use and think about calculus. Happily, dissatisfaction with the traditional calculus is near universal today, there are very few doubting Thomases. This welcome crisis was brought on by the widespread availability of powerful pocket calculators that can integrate functions, find their maxima, minima, and zeroes, and solve differential equations with the greatest ease, exposing the foolishness of devoting the bulk of the calculus course to antiquated techniques that perform these tasks much more poorly or not at all. We now have the opportunity to sweep clean all the cobwebs and dead material that clutter up calculus. We have to think carefully what we put in its place; I strongly believe that calculus is the natural vehicle for introducing applications, and that it is applications that give proper shape to calculus, showing how and to what end calculus is used...No doubt computing will play a large role in undergraduate education; just what will take a great deal of experimentation to decide. The brightest promise of computing is that it enables students to take a more active part in their eduation than ever before."

In *Calculus with Applications and Computing*, freshman calculus was freshly reconceived. Peter wanted to bring the course closer to the needs and interests of practitioners in science and engineering and to take advantage of the teaching possibilities of high-speed computers. By easy computer programs, interesting, significant applications became accessible, which without computer assistance would be forbiddingly tedious. Students of this book were assumed to be simultaneously learning computer programming and ready to apply their new computer skills in the calculus course.

Peter wrote in his introduction, "The educational value of numerical examples worked out by the student themselves, individually or as members of small teams, cannot be over-estimated. A good student is likely to be a better and more enterprising computer programmer than his instructor, and this will enable him to experiment on his own instead of being bound to his text or to the instructor's apron strings.

"Finding numerical answers is a very important part of an application of mathematics...the calculation of a well-chosen special case can take the role of a crucial experiment in confirming an old speculation or pointing to a new one. We do not hesitate to break with tradition where we feel that a change is called for." In particular, the explicit integration of special elementary functions, which is of minor interest to practitioners, is dealt with briefly.

Real numbers are seen in *three-dimensional* different disguises: as algebraic entities, as points on the number line, and as infinite decimals, along with a daring philosophical insight: "Infinite decimals are Platonic ideals of which we mortals only see shadows that appear as finite digits on the register, or printout, of calculation."

As a payoff for defining real numbers as infinite decimals, "The notion of a convergent sequence can be explained without resort to Greek letters [Epsilon and delta are being snubbed!!] merely by noting that a sequence converges if more and more of the digits of its members are identical."

Still more radical, the exponential function is *defined* as a model for growth of bacteria and decay of radioactive elements. Simple physical reasoning about both models then easily shows that, as a function of time, the amount of radioactive material, or the bacterial population, satisfies an equation:

$$M(s+t) = M(s)M(t).$$

From this formula one easily and rigorously obtains the differential equation of the exponential function, its local and asymptotic behaviors, and those of its inverse, the logarithm.

Peter had long held that the standard notion of "convergence of a sequence of functions", pointwise convergence, is the wrong one to teach. A stronger kind of convergence, "uniform convergence", is more useful. Although tradition reserves it for an advanced course, Peter uses it right from the beginning. He even writes, "We define uniform convergence of a sequence of functions, a natural, useful, and elementary concept that Victorian prudishness usually reserves for mature audiences."

Peter states his educational credo firmly: "The authors of this text feel that not enough mathematical talent is devoted to furthering the interaction of mathematics with other sciences and disciplines. This imbalance is harmful to both mathematics and its users; to redress this imbalance is an educational task which must start at the beginning of the college curriculum. No course is more suited for this than the calculus; there students can learn at first hand that mathematics is the language in which scientific ideas can be precisely formulated, that science is a source of mathematical ideas which profoundly shape the development of mathematics, and last but not least that mathematics can furnish brilliant answers to important scientific problems."

As it turned out, the assumption that calculus students were learning simple computer programming is not the reality of American college education. "Our calculus book was enormously unsuccessful, in spite of containing many excellent ideas. Part of the reason was that certain materials were not presented in a fashion that students could absorb. A calculus book has to be fine-tuned, and I didn't have the patience for it. Anneli would have had it, but I bullied her too much, I am afraid. Of course, there has been a calculus reform movement and some good books have come out of it, but I don't think they are the answer. First of all, the books are too thick, often more than 1,000 pages. It's unfair to put such a book into the hands of an unsuspecting student who can barely carry it."

So Peter now has worked with Maria Terrell, a professor of mathematics specializing in calculus at Cornell University, to create a new book, *Calculus with Applications*. Computing has been omitted from the title but survives in the text.

One of the pleasures of writing this book is the opportunity to quote from Peter's 1977 article, "The bomb, Sputnik, computers, and European mathematicians".

"No chronicle of the fifties is complete without a mention of the McCarthy era and its effect on science. Here are a few highlights: In 1950 the Regents of the University of California imposed a loyalty oath on its faculty. As I recall the oath was innocuous in itself; but the idea that the Regents could impose it on professors was odious. Independent minded faculty members, among them a number of scientists and mathematicians, quit rather than sign. They were fully vindicated when the courts declared the oath illegal. One can take particular satisfaction in the fact that today's President of the University of California was one of the nonsigners 25 years ago. A similar loyalty oath requirement was imposed in Oklahoma, where it resulted in Aronszajn moving from Stillwater to Lawrence, Kansas. The campaigns of McCarthy and of the House Un-American Activities Committee were a serious threat to the scientific community. Although there was very little sympathy for hard core members of the Communist party who have followed the party line through every twist and turn and served as apologists for Stalin's byzantine cruelties, scientists realized that academic freedom is indivisible, and that the aim of the charges was not to reveal some nonexisting conspiracy, but to make political hay for the fearless vampire hunters. I never had to prove my loyalty in a formal security hearing, but several of my friends did. In two cases I testified for the defense, in both instances the charges were flimsy, but at least they were stated above board and—being flimsy—were dismissed. Other victims had to fight vague, Kafkaesque charges, based on evidence in closed files. In those days the nation sadly lacked a Freedom of Information Act. By and large, universities and especially individual mathematicians behaved honorably. We closed ranks, stood up to our tormentors, defended the unjustly accused, and scrambled to find jobs for those who were unfairly dismissed. So the long-term damage was small."

In addition to working at NYU and for the U.S. government, Peter was vice president of the AMS from 1969 to 1971 and president from 1977 to 1980.

In my first sabbatical leave in 1971 I was back at Courant as a postdoc, invited by Peter Lax. I worked with George Papanicolaou on random evolutions, and with Martin Davis on two more *Scientific American* articles, one on nonstandard analysis, one on Hilbert's tenth problem. (The latter won a Chauvenet Prize from the MAA.) At my next sabbatical, in 1979, I was working with Phil Davis at Brown University on our book *The Mathematical Experience.* Peter gave the helpful suggestion to question the epistemological status of the theory of finite simple groups. By great good luck Anneli Lax was friendly with the editors of the *New York Review of Books*, Barbara and Jason Epstein. Anneli suggested reviewing our book, and they assigned it to the famous math buff Martin Gardner, who gave it a mostly favorable review.

From 1980 to 1986 Peter was chairman of the National Science Board, the governing body of the National Science Foundation.

"Being on the National Science Board was my most pleasant administrative experience. I found out what making policy means. Most of the time it just means nodding 'yes', and a few times saying 'no'. But there are sometimes windows of opportunity, and the Lax Panel was a response to such an opportunity."

The most remembered action Peter took on the National Science Board is generally referred to as the "Lax Report". When Cray's CDC 7600 came out, no university could afford to buy one. University scientists couldn't use one unless they had a contact at Los Alamos or Livermore or the Center of Atmospheric Research in Colorado. "That was an intolerable situation. My friends who were interested in large scale computing—in particular, Paul Garabedian—were complaining about it. The government alone had enough money to purchase these supercomputers, and it had stopped placing them at universities. They went only to national labs and industrial labs. Unless you happened to have a friend there, you had no access. That was very bad for computational science, because the most talented people were at the universities. But accessing and computing at remote sites was made possible by the ARPANET, which would become the model for the Internet.

"The idea for a report came from Kent Curtis, who was head of computer science at NSF, and his boss Jim Infante strongly supported it. We made a stink in issuing the report in 1982, and there was action on it. The panel that I established made a strong recommendation that the NSF establish computing centers. My comment on our success was to misquote Emerson: 'Nothing can resist the force of an idea that is five years overdue.' By 1985 we had supercomputer centers at a number of universities up and running, funded by NSF."

In his interview for the SIAM archive, Colella asked Peter, "How did you make the argument? What was the justification?" Peter answered, "Lawyers have a neat Latin expression, 'Res ipsa loquitur' (the case speaks for itself). We added the obvious argument: the most creative scientists are at universities, and if they don't have access to supercomputers, they will only do things that don't involve supercomputers. I remember a study of the speed of numerically solving the discretized Laplace equation. The speed had increased by some factor, call it 'n', over a period of thirty years. One square root of that factor of increase was due to increased machine speed, the other square root was due to improved numerical methods."

Colella informed Peter that the current value of the square root of "n" was actually about 16 million.

Lax: Very impressive. In the halcyon period, all government agencies realized how vital scientists are, for defense and everything else that affects national life. In those days, the administrators had a vision of their own. Von Neumann was one of them, on a very high level! A good, understanding administrator plays a very important role."

In 1982 he was elected to the U.S.A. National Academy of Sciences and also to the Paris Academy of Sciences. In the same year he addressed the International Congress of Mathematicians and was granted the National Academy of Sciences Award in Applied Mathematics and Numerical Sciences. In 1986 he received the National Medal of Science. That year was his sixtieth birthday. In his honor there was a week-long special series of research lectures at Stanford University. (This seems to be the obligatory format for mathematical birthday parties.) In 1986 I was present in the audience.

In 1987, jointly with Kyoshi Ito, the creator of the stochastic calculus, Peter Lax won the Wolf Prize of Israel, receiving a diploma and $100,000 cash. This prize was established in 1976 by a gift of $10 million from Dr. Ricardo Subirana y Lobo Wolf (1887–1981), a Cuban Jewish inventor, diplomat, and philanthropist, and his wife, Francisca (1900–1981). Five or six Wolf Prizes are awarded yearly to outstanding scientists and artists "for achievements in the interest of mankind and friendly relations among peoples." The scientific fields are agriculture, chemistry, mathematics, medicine, and physics.

The citation reads, "Professor Peter D. Lax, a graduate of the Courant Institute, embodies the best traditions of D. Hilbert as continued by R. Courant. Among his many contributions are the solution of the Cauchy problem with oscillatory data, the clarification of the role of stability of a numerical scheme, the comprehensive development of scattering theory, the theory of non-linear conservation laws and a deep insight into the Korteweg-de Vries equation. Prof. Lax's influence has been profound and decisive in both pure and applied mathematics."

CHAPTER 6. LATER CAREER

In the eighties I was a member of a Mathematicians' Action Group that distributed antiwar leaflets and demanded the floor, year after year, at national meetings of the AMS. Two founders of that group, Lee Lorch and Chandler Davis, were old friends of mine. Peter Lax disagreed with the MAG demand that the AMS end cooperation with the U.S. military. In 1988, in "The flowering of applied mathematics in America" (*A Century of Mathematics*, p. 462), after summarizing U.S. military support for mathematics since World War II, he wrote, "In view of the distinguished past and present success of this research program, it came as an utter surprise that a group within the AMS proposed to reduce support for mathematics by the DOD. Many who were supported by the DOD were deeply offended by the suggestion that they were accepting money from a tainted source, and that the support should have been given to worthier recipients." I know that our objection to military support for mathematics did not have financial envy as a motive.

In 1988, by falling in love, I unwittingly served to revive Peter's connection to prewar Budapest. Forty years earlier, Dr. Henry Lax in New York had received a letter from Dr. Ferenc Polgár in Geneva, Switzerland. His old colleague in radiology at the Jewish Hospital in Budapest had survived the war with his family! Now Dr. Polgár was hoping to bring them to the U.S. and was asking, would Henry Lax "sponsor" them—guarantee that they would not become charges on the charity of the U.S. government, as the playwright Ferenc Molnar a few years before had guaranteed that the Laxes would not become such charges? Of course, Henry was glad to help. He and Polgár had actually been classmates in medical school in Budapest.

The Polgárs came to the U.S. Dr. Polgár's daughter, Vera John-Steiner, became a psycholinguist and my colleague at the University of New Mexico. While writing her book *Notebooks of the Mind*, she interviewed me in the mathematics department of UNM. From then on I cherished our all-too-infrequent conversations. Subsequently, we became a couple.

I don't remember when we first realized that the immigration sponsor of Veronka's family was the father of my mathematical mentor. It was a great occasion when she and I first went together to the apartment on the ninth floor of The Eldorado. Later, Veronka and I wrote an article on "the Hungarian miracle", the great number of brilliant Hungarian mathematicians of the time between the Great Wars. Peter gave us a lot of help. When he comes to New Mexico to give talks and visit friends, it is a welcome chance for Peter and Veronka to speak their native tongue.

In 1992 Peter received the Leroy P. Steele Prize of the American Mathematical Society "for his numerous and fundamental contributions to the theory and applications of linear and nonlinear partial differential equations and functional analysis, for his leadership in the development of computational and applied mathematics, and for his extraordinary impact as a teacher."

In 1996 Peter had his seventieth birthday. Since Louis Nirenberg's seventieth birthday had been only one year previously, there were joint birthday celebrations for the two friends in the form of research meetings and seminars in Venice, Italy, in June and at the Villa La Pietra in Florence, Italy, in July. His eightieth birthday was celebrated along with Nirenberg's at the Courant Institute in October 2005 and at the Universidade de Rio de Janeiro in Brazil in May 2010. (I participated there.)

From the mid-1950's to the mid-1990's, Peter published around one hundred thirty research papers, around three a year on average. There were sequences of papers, deepening and strengthening certain continuing themes: shock waves, conservation laws, finite difference approximations, solitons, and the on-going collaboration with Ralph Phillips on scattering theory. Along with Peter's twofold sequence of doctoral students, in either computing or in theory, there is a twofold sequence of publications. Around one fourth of his publications are algorithmically or computationally oriented.

In addition to the books and the research articles, there are scattered works of exposition, history, biography, book reviews, and memoirs. Peter takes justifiable pride in writing well. His biographical articles on John von Neumann and Richard Courant deserve to be accessible to the general reader and are included in this book as appendices.

I have already mentioned the books that came in the 1960s and 1970s: *Scattering Theory*, with Ralph Phillips, in 1967 and revised in 1989; *Hyperbolic Systems of Conservation Laws and the Mathematical Theory of Shock Waves* in 1973; and in 1976 two books, *Scattering Theory for Automorphic Functions*, with Ralph Phillips, and *Calculus with Applications and Computing*, co-authored with Sam Burstein and Anneli Lax. Chapter 8, following here, is about five books he published between 1997 and 2012: *Linear Algebra* in 1997, *Functional Analysis* in 2002, *Selected Papers* in 2005, *Hyperbolic Partial Differential Equations* in 2006, and *Complex Proofs of Real Theorems* with Lawrence Zalcman in 2012.

Teaching at the Courant Institute of New York University descends from an illustrious tradition that goes back to Felix Klein (1849–1925). Klein began as a wunderkind professor at Erlangen at age twenty-three. He competed intensely with the French mathematician Henri Poincaré in developing the theory of automorphic functions. In his full maturity, Klein became the éminence grise of German mathematics. He thought Germany needed a school of mathematics oriented toward science and industry, in contrast to mathematics in Berlin, where Professors Frobenius, Kummer, Kronecker, and Weierstrass cherished purity and rigor above all.

With the leadership of David Hilbert, who was open to ideas and problems from everywhere, Klein's goal was achieved at Göttingen. Göttingen attracted brilliant visitors and students from Europe and around the world. Among the epoch-making achievements at Göttingen were the foundations of quantum mechanics by Born and Heisenberg, the work on the foundations of logic and set theory by Hilbert and Bernays, the creation of abstract algebra led by Emmy Noether and Emil Artin, and the development of the mathematical tools of modern physics in the works of Hilbert and his leading associate Richard Courant.

Unlike the traditional lecture style, where perfectly prepared notes are copied onto the blackboard by the professor to be then copied from the blackboard by the students, Hilbert often got stuck or confused in his lectures. He actually thought through and re-created the mathematics there in front of the class! For students who want to become creative mathematicians themselves, such imperfect lectures are far more instructive than letter-perfect prepackaged ones, such as the lectures delivered by Herglotz. Richard Courant too was a lecturer who let his students observe and participate in his own creative thinking. This style of teaching, which

I enjoyed as a student of Peter Lax, is an important part of the heritage of the Courant Institute.

Peter has supervised 55 Ph.D. students and has 518 mathematical descendents. The names of his doctoral students are listed in Appendix 6. Most of them became creative mathematicians. But Peter admits that he sometimes fails to recognize former students. He says, "I have poor face recognition. Once I was invited to give a talk at a university in California. There was tea with the faculty before the talk, an occasion for conversation. The safest opening on such occasions is 'What are you working on?' The fellow I asked named a topic; I replied that I have never worked on this subject, but once had a student who wrote his dissertation on it. 'I am that student!' he answered indignantly."

Looking over the list of his students, one easily classifies them as either theoretical or computational. Some did a computational thesis and had a career in scientific computation, becoming heads of large staffs and even members of the National Academy. Others did a theoretical thesis and had careers teaching pure mathematics and publishing in journals of pure math. It is hard to find one who does not clearly fall into one group or the other. I was theoretically oriented; I managed all my life to avoid doing numerical work. Peter suggested using the idea of my thesis to prove stability for the corresponding difference approximations, and I did write a paper following his advice. But I didn't do a single numerical example.

Besides the followup on my doctoral thesis suggested by Peter, at the age of fifty I published a second paper on finite difference approximation, virtually my last piece of mathematical research. Tosio Kato was a co-author. I am rather proud of it; I think it deserves to be remembered and referred to.

Both of these papers are rigorously theoretical. In both of them it is carefully proved that a certain class of difference approximations actually converges and is stable. In the second, later piece an explicit method is given to achieve accuracy of *arbitrarily high order* for *any* well-posed linear evolution equation—hyperbolic, parabolic, or whatever. This was and is quite surprising—when first told of it, several experts were incredulous.

Yet neither paper contains a single numerical computation! The job of writing code and running the program is simply ignored. I didn't follow up on my own ideas when they showed promise of utility in practical computation. I didn't know how to do it, and I wasn't interested in learning how. Only now, so many years later, do I see how strange this is.

In fact, as I now understand, my failure to make any attempt to "cash in" to reach the stage of concrete utility or usefulness resulted from a combination of two of my attitudes or values. I found computer coding or even just learning a computer language utterly unappealing. And politically, I wouldn't get close to anything involving security clearance or military applications.

CHAPTER 7

The queen of Norway

Mathematics organizations love to give prizes. At the annual joint meeting of the American Mathematical Society and the Mathematical Association of America, hours are spent awarding the Chauvenet Prize, the Ford Prize, the Steele Prizes, and so forth and so on. On the other hand, recipients of prizes generally don't seem to feel that getting a prize is much of a big deal. In 1968 the Birkhoff Prize for Applied Math (named after the elder Birkhoff, George David) was split three ways between Garrett Birkhoff, Clifford Truesdell, and Mark Kac. I happened to eat lunch with Peter Lax that day. He thought it a waste to give another prize to Kac. "He already has received plenty of prizes. Better to give it to someone younger and underappreciated, like Cliff Gardner."

Peter receiving Abel Prize from Norwegian H.R.H. Crown Prince Haakon in University Aula, Oslo, 2005

Peter's grandsons Pierre, Tommy and Timmy at the Abel Prize ceremony

Peter Lax was elected to nine academies of science and honorary societies, starting with four in 1982 (Paris, U.S. National, American, and New York) and going on to the Russian, Hungarian, Beijing, Moscow, and London.

He received eleven honorary doctorates, beginning at Kent State University in 1975 and Paris in 1979 and going on to Pennsylvania and Tulane Universities in 2012.

Because Alfred Nobel did not deem mathematics worthy of his prize, the biggest prize in mathematics was traditionally the Fields Medal. It is awarded only once every four years at the International Congress of Mathematicians. It can be given to up to four people—they must be under the age of forty. It is named for the Canadian mathematician John Charles Fields, who was instrumental in establishing the award, designing the medal, and funding the award, now $15,000 Canadian dollars (roughly US $15,000). Peter Lax did not get a Fields Medal. This might be, as he suggested, because the Fields Medal has often recognized the solution of a famous problem, and he is known rather for creating or discovering problems and then solving them. The Fields Medal has never been awarded for work in applied mathematics.

For over a century it was proposed that Norway set up an Abel Prize in mathematics to fill the gap left by the absence of a mathematics Nobel Prize. The mathematician Niels Henrik Abel (1802–1829) was Norway's greatest scientist. At age nineteen he solved a problem that had been studied without success for 250 years. He showed that there is no solution for the roots of a quintic equation or any general polynomial equation of degree greater than four in terms of roots. To do this, he invented (independently of Galois) the branch of mathematics known as

Peter with KappAbel student competition winners, 2005

group theory. Among his other accomplishments, Abel wrote a monumental work on elliptic functions which was published after his death.

In 2000 Nobel's oversight was corrected when the Norwegian Academy of Science established the Abel Prize as an international prize for outstanding work in mathematics, including mathematical aspects of computer science, mathematical physics, probability, numerical analysis and scientific computing, statistics, and applications of mathematics in the sciences. The prize recognizes contributions of extraordinary depth and influence which may have resolved fundamental problems, created powerful new techniques, introduced unifying principles, or opened up major new fields of research.

The Abel Prize is much more lucrative than the Fields Medal—6 million Norwegian kroner (about 750,000 euros). Like the Nobel Prize, it is Scandinavian in location but international in scope. Its awardees have been at the highest pinnacle of mathematical achievement and prestige. It does not carry any inconvenient age limitation and has been given to mathematicians in the fullness of their years after their great work had survived decades of continued admiration and appreciation.

The first proponent of an Abel Prize was Norway's other world-famous mathematician, Sophus Lie. Before his death in 1899 he was gathering support for an Abel Prize in pure mathematics to be awarded once every five years. There was overwhelming support from leading European mathematical centers, but when Lie died the effort died too.

In 1902 King Oscar II of Sweden and Norway became interested in a prize in Abel's honour. The mathematicians Carl Størmer and Ludvig Sylow drew up rules for such a prize. However, when the union between Sweden and Norway was dissolved in 1905, the project had to be abandoned. Norway was not then in a financial position to establish an Abel fund on its own.

In August 2000 Abel's biographer, Arild Stubhaug, discussed these events from a century earlier with Tormod Hermansen, the CEO of Telenor Group, the Norwegian telecommunications company (an international wireless carrier with 203 million subscribers in Scandinavia, Eastern Europe, and Asia as of 2010). Hermansen briefed the Norwegian Ministry of Education, Research and Church Affairs about the idea of an Abel Prize, and Stubhaug took it to the University of Oslo's Department of Mathematics. On May 23, 2001, a working group from the Department of Mathematics sent Prime Minister Jens Stoltenberg a proposal for an Abel Prize. On August 23, 2001, the prime minister announced the establishment of the Niels Henrik Abel Memorial Fund with 200 million Norwegian kroner to award each year the Abel Prize for outstanding work in mathematics.

The Norwegian Academy of Science and Letters appoints five outstanding research scientists in mathematics to be each year's Abel Committee. That committee recommends a candidate to the academy, which announces the name of the new Abel Laureate in late March or early April. The prize was first awarded on June 3, 2003, to Jean-Pierre Serre of the Collège de France.

In 2005 Peter Lax received the third Abel Prize, following Michael F. Atiyah of the University of Edinburgh and Isadore M. Singer of MIT jointly in 2004. He was followed by two of his colleagues at the Courant Institute, the probabilist S. R. Srinivasa Varadhan in 2007 and the geometer Mikhail Gromov in 2009.

The Abel ceremony was a great occasion for Peter and Lori and their families.

I asked Marnie, "How did Peter first hear about winning the prize?"

Marnie: It was like the secrecy of the Nobel Prize. Jimmy called me and said, "Something is happening, but I can't tell you." Then later we heard the announcement, and Jimmy called and said, "Can you come?" "Can we come? What, are you kidding?"

Peter and Lori were already in Oslo. Going to Oslo were Jimmy's three boys, Tommy, Timmy, Pierre; and Helen, Anneli's brother's daughter, and her husband, Michael, from California; and Danny and Cynthia—Danny is Jimmy's best friend; they've known each other since—and then Lori's grandniece, Katherine Courant; and Lori's daughter, Suzy, and her son, David.

Reuben: And I know Burt Wendroff was already in Norway. His wife is Norwegian.

Marnie: Once we got there, everything was taken care of. Peter had a handler, a very lovely woman who made sure he arrived here and there. They had a limousine. The government did all of this. It was incredible. There were all these various dinners, lunches, and receptions, all on the largesse. And it was May. Lilacs were blooming.

David: I love lilacs...they were all different colors, they were huge, all over downtown Oslo.

Marnie: Pierre must have been maybe twelve, and Timmy was probably in high school. They just had a great time. They were so proud of Peter, and they got to be there with their brothers and with Jimmy.

"We flew up to Bergen. In Bergen they had a fair for elementary school kids. It was pouring rain. I mean buckets of rain. And these kids had tents; speakers would go in and give a talk to them. They were all adorable in their little foul-weather gear jackets and their boots. And Peter drives up in this huge white stretch limousine, and the kids went nuts! They were all running out of their tents and squealing after this car like he's a rock star! I'm not so sure if it was because of Peter or because of this car...it was so grand.

Reuben: Did you meet the queen?

Marnie: Very charming, very attractive. There's a wonderful picture of Jimmy with her. I was too embarrassed to ask to have my picture taken...silly...

David: Wasn't the prince very egalitarian too? He'd put the chauffeur in back and drive occasionally.

Marnie: We went to a high school where they gave a play acting out Abel's life...very sweet.

Reuben: A tragic life. He died just before he became famous.

Marnie: Yes. The makeup of the students was so diverse. They weren't just Scandinavian. There were all sorts of ethnicities. The prize ceremony was in a beautiful stone building. On the walls was a circumferential mural by Munch. And they had these beautiful opera singers. It was just unreal. There was awe all around us...

Reuben: And Peter made a speech?

Marnie: It was a very short, simple speech. He was very humble. You know how he sort of blushes when he talks. He was clearly overcome.

Reuben: And then you had a dinner?

Marnie: We had so many...there was one dinner with the queen; it was a banquet.

David: Two hundred people. Your usual royal banquet.

Marnie: So we all came bustling back. I almost couldn't go, but managed to wangle the time off.

In an article celebrating Peter's receipt of the Abel Prize, Helge Holden of the Norwegian University of Science and Technology in Trondheim wrote, "The Prize was awarded to Lax for his ground-breaking contributions to the theory and application of partial differential equations and to the computation of their solutions. The difficulty in giving a description of Lax's contributions is that they are so numerous and important that in an article of this length it is impossible to do them justice."

Entr'acte. Peter's stories

As I have already mentioned, Peter is known among his friends as a storyteller. I was able to include several of Peter's stories in the chapter about Los Alamos. The ones that do not naturally fit into Peter's life story are presented here. Each little story is self-contained. The point or the moral is left for the reader to detect. The order can be called "almost pseudorandom". To keep the right flavor, the stories are left in the first person of Peter Lax. In the following stories, "I" means Lax.

****

In 1959 there was a committee meeting at Convair, an airplane and missile manufacturer, concerning ICBM's (intercontinental ballistic missiles). The head of the German ballistic missile project during the Second World War was Werner von Braun. One of his associates was hired by Convair, and he made a presentation, using the image of dropping a missile into a pickle barrel in Red Square. Theodore von Kármán expressed his skepticism by telling the story of the rabbi of Lemberg, whose disciples claimed he could see events far away. For example, he saw Kracow in flames. When a skeptic objected that Kracow was not in flames, the disciple replied, "True, but isn't it remarkable that he sees as far as Kracow at all?" Von Kármán concluded, "Isn't it remarkable that he sees as far as a pickle barrel in Red Square?"

****

Otto Neugebauer was an illustrious historian of mathematics, an émigré to the U.S. from Germany. Corresponding with a friend in Germany who had asked why he did not use their *Muttersprache*, he explained that "the language of my correspondence does not depend on my mother's language, but on my secretary's." Neugebauer said that the European Gymnasium prepares you better for life than an American high school because "you learn who your enemy is—the teachers."

****

Antoni Zygmund, the great Polish Fourier analyst who became a professor at the University of Chicago, had only one comment on the great American sport baseball: "The World Series should be called the World Sequence."

****

Both Peter and his wife, Anneli, were members of the Courant Institute faculty, as were the two sons-in-law of Richard Courant. To a visitor perplexed by this, Joe Keller explained that the nepotism rule at the institute is that each professor may have one relative on the faculty and Richard Courant, two.

****

The famous English mathematician George Neville Watson and the German refugee physicist Rudolf Peierls were colleagues at the University of Birmingham. Watson offered to assist Peierls with his mathematical problems under one condition: that Peierls would not explain where the problem comes from.

****

The next story needs some background: Mark Krein was an eminent mathematician living in Kiev, a Jew, and the Ukrainians have a certain history regarding Jews. A certain U.S. mathematician started his lecture at the Ukrainian Academy of Sciences by saying how honored he was to address the scientific academy which has the highest standards in the world. When asked to explain, he said, "An academy of which Mark Krein is only a corresponding member has the highest standards in the world."

****

Helmut Hasse was a leading number theorist who became the head of the institute at Göttingen after Courant was kicked out by the Nazis. Natascha Artin, the wife of the great algebraist Emil Artin, Emmy Noether's disciple, was half Jewish. Hasse tried to convince Artin not to emigrate to the U.S. by offering to have the couple's one-quarter Jewish children declared honorary Aryans. He made no offer about Artin's half-Jewish wife. Before the triumph of Hitler, Hasse had taken pride that one of his ancestors was the great Jewish composer Felix Mendelsohn. His anti-Nazi colleague Carl Ludwig Siegel always referred to him as Herr Hasse-Mendelsohn.

****

Hermann Amandus Schwarz was an eminent professor in Berlin in the nineteenth century. He liked to say: "I married the daughter of my esteemed teacher Kummer. I would rather have married the daughter of my even more esteemed teacher Weierstrass, but unfortunately he had no daughter." (Weierstrass was a bachelor.)

****

At a mathematical meeting in St. Louis a jazz singer provided entertainment. Before singing the "Empty Bed Blues" she explained, "You mathematicians should realize that an empty bed is a bed with one person in it."

****

At the end of World War I a group of journalists had a competition for the most sensational headline. The winning entry was: "Franz Ferdinand Found Alive, World War Fought by Mistake."

****

The mathematician Andre Bloch suffered a head injury in World War I, which made him subject to fits. In one of his fits he killed several members of his family. He was put away in an insane asylum, where he continued to work on mathematics. An enterprising journalist interviewed him about his life. The last question he asked was what mathematical problem he was working on. Bloch replied that he was working on a problem in projective geometry that involved the two imaginary points at infinity that lie on every real circle. The journalist added in his write-up, "What a tragedy that such a brilliant mind is clouded forever."

****

Heisenberg was driving a car with a passenger and scared the passenger with his high speed. At the end of the trip, the passenger demanded, "Did you realize how fast you were going?" Heisenberg replied, "No, but I knew exactly where I was."

****

The Texas topologist R. L. Moore asked his class to give a counterexample to the false theorem "Every bounded set of numbers has a largest element". A student proposed the following counterexample: "The set of two elements, the numbers 0 and 1". Being a set of only two elements, it has a *larger* but not a *largest*.

****

Cathleen Morawetz's husband, Herbert, referred to Emil Artin as one of the greatest of mathematicians. But Cathleen corrected him; he protested, "That was what you yourself said." She answered, "Oh, that was when he was alive; he was one of the greatest living mathematicians. Now he is dead, and he isn't one of the greatest among those."

****

When the Wolf Prize of Israel was awarded for the first time, Peter was on the award committee and thought the prize should go to Carl Ludwig Siegel. But the other committee members insisted on Israel Moiseyevich Gelfand, so finally it was agreed to share the prize between Siegel and Gelfand. Peter was gratified when Gelfand said, "It was a great honor for me to share a prize with Siegel."

****

Lipot Fejer was one of the first of the famous Hungarian mathematicians of the twentieth century. Speaking mathematically about real life, he said that a professor's salary is necessary but not sufficient.

****

In 1958 Lipman Bers of NYU was the first U.S. mathematician invited to the Soviet Union after the world war. Petrovsky, rector of Moscow University, threw a party in his honor. Israel Gelfand, then about forty-five years old, explained to Bers that progress in mathematics depends not so much on the discoveries of the brilliant young mathematicians, but on the older mathematicians who have a vast knowledge and have a sense of which direction is open for progress. Bers was intrigued, but then Petrovsky remarked, "Professor Gelfand did not always hold this view."

****

At a meeting Harish-Chandra asked me, "Why is Hermann Weyl regarded as such a great mathematician? Everything he did was rather easy." I did not venture a reply, but some time later I was with Atle Selberg and repeated this remark to him. "You see," he said, "there are two kinds of great mathematician. The first kind solves a difficult problem; it becomes thereafter easy. Such a mathematician was Hermann Weyl. The second kind solves a difficult problem, it still remains difficult; such a mathematician was Carl Ludwig Siegel." I didn't have the courage to ask him to which class he himself belonged.

****

Friedrichs, while visiting Germany in the 1960s, met Heisenberg and couldn't resist expressing to him the gratitude of mathematicians for creating quantum theory, which gave rise to so much interesting mathematics. Heisenberg allowed that this was so. Friedrichs ventured to add that, to some extent, mathematicians have repaid this debt. Heisenberg was noncommittal, so Friedrichs said, "After all, it was a mathematician, von Neumann, who clarified the difference between a self-adjoint operator and one that is merely symmetric." "What's the difference?" asked Heisenberg.

****

After I was elected to the National Academy of Sciences, I wanted to propose Friedrichs to become a member, and I asked Paul Cohen to support my proposal. But Paul said, "What hard problem did he ever solve?" I pointed out that Friedrichs's theory of symmetric hyperbolic equations was a great success, but Paul was not moved. Nevertheless, after some more effort, Friedrichs was elected to the academy. When I gave him the news, he commented, "You know, I never did solve a really hard problem."

****

Enrico Bombieri was proposed for a professorship at the Institute for Advanced Study in Princeton. At a faculty meeting somebody presented his brilliant work on the calculus of variations, on minimal surfaces, on algebraic number theory. Everybody was very enthusiastic except Andre Weil, who said, "This man Bombieri, he dabbles in this, he dabbles in that, he will end up like von Neumann."

****

Enrico Fermi said, "Edward Teller is a monomaniac with many manias."

****

The favorite day of the year at the Hoover Institute in Palo Alto is daylight saving time, when you turn the clock back.

****

At a dinner Monroe Donsker asked his neighbor, an astrophysicist, about man-made satellites. He replied, "Personally, I have never worked on anything in our galaxy."

****

Von Neumann got stuck in a lecture. He said, "I knew three ways to prove this theorem, but unfortunately I chose a fourth one."

****

Warren Ambrose had a job offer from the University of Chicago and then received one from MIT. He called Marshall Stone, the chairman at Chicago, to discuss the matter. Stone said, "Let me make it easy for you. I take back my offer."

****

Women mathematicians usually have fathers in math or related fields. I asked Dorothy Bernstein what her father was. "A farmer." How did you get into math? "From being a student of astronomy." Why did you choose astronomy? "So I would have no curfew at the dorm."

****

Kakutani was a Japanese citizen residing in the U.S. After Pearl Harbor, he was repatriated to Japan. While on the boat he found a beautiful theorem. Not sure if he would arrive safely, he wrote it out, put it into a bottle, and threw it into the sea.

****

Before being sent out of the U.S., Kakutani went on a walk with Paul Erdős. They wandered into a restricted area. They were picked up by police as suspicious foreigners and asked what they were doing. Erdős explained, "We were trying to prove a theorem."

****

Some idioms of the English language create difficulties for those who learned a different tongue in their mother's arms. Once Peter needed to ask some mathematical question of his honored mentor, Kurt Otto Friedrichs, and said it was a topic of which Friedrichs was "past master". Friedrichs thought Peter was calling him a "has-been". Apology and explanation were of course offered and accepted.

****

George David Birkhoff was a leading U.S. mathematician in the first third of the twentieth century, remembered today both for great research in dynamical systems and for keeping his department at Harvard free of Jewish faculty members. However, he did not necessarily refuse to recommend Jewish candidates for departments other than Harvard's. Peter heard the following story from someone who was actually present. Birkhoff was talking on the telephone to the chair of the math department at the University of Rochester, where a Jewish refugee had not received the job offer that Birkhoff felt Rochester should make. It seems Birkhoff assumed, rightly or wrongly, that Rochester's failure to come through with the expected offer was an expression of anti-Semitic bias, for he was heard shouting over the phone at the Rochester math chairman, "Who do you think you are, Harvard?"

****

How did the eccentric crew of European mathematical immigrants fare in the land of the Puritans? Most of them were eager to assimilate. An extremely distinguished mathematician of German origin who settled in Princeton went so far as to decide to master baseball, and set aside Saturday afternoons for that purpose. A visitor, calling during the sacred hour, was warned at the door by the great man's wife, "Ssh! Hermann is listening to the ball game."

****

Other immigrants tried to cling to the ways of the Old World. One of them was heard explaining to a friend, also from Europe, "Now that we are in America we must behave as Americans, so in public you must call me Stefan and I will call you Hilda. But of course when we are among ourselves you will continue calling me Herr Professor and I will call you Frau Professor Doktor." Such an attitude was exceptional and temporary; eventually all refugees became as American as Apfelstrudel.

****

I would like to recall here an incident which has partly sinister, partly comic overtones, concerning the appointment of Norman Levinson at MIT. Hardy was visiting Cambridge when the matter was proposed, and being an admirer of Levinson, he gave his endorsement in person to a high administrator. He was told there was difficulty making the appointment because there was no room for another Jew on the MIT faculty. Hardy said he didn't realize that MIT was an Institute of Theology, and threatened to expose the affair in the pages of *Nature* unless the appointment went through, and it duly did.

****

(Here I (R.H.) must post an addendum to Peter's story. In the memoir of Levinson's widow, Fagi, it is explained that the Jewish mathematician Jesse Douglas, then a faculty member at MIT, was being dismissed because of difficulties in his classroom performance (Zipporah (Fagi) Levinson, *Recountings*, pp. 1–22, A K Peters, Ltd., Wellesley, MA, 2009). Thus, the appointment of Levinson in fact did not increase the number of Jews on the MIT faculty, but merely prevented it from decreasing. Nevertheless, Hardy may well have been correct in his belief that his threat was needed in order to push through the appointment.)

CHAPTER 8

Books

All Peter's writings share a consistent philosophical and methodological viewpoint toward research, education, and applications. It is the viewpoint of the Courant Institute and of Richard Courant, David Hilbert, and Felix Klein. This philosophical context is not a matter of metaphysics, ontology, or epistemology; it's a matter of the proper goals and methods of mathematical life and work. I will concentrate on the conceptual rather than on the technical and computational aspects of these books. In writing about some of them, I assume some background knowledge. Choose what to read and what to skip.

The subject of linear algebra became an essential part of an undergraduate math degree shortly before I started graduate school in 1957. I had been out of school for ten years and had to learn it from scratch as a graduate student. Fortunately, NYU then offered a full-year beginning graduate course in linear algebra from the bottom up. Since those days, there has been a great outpouring of undergraduate linear algebra texts. Many of them are abstract and set-theoretical. Some of them follow Paul Halmos's pioneering *Finite-Dimensional Vector Spaces* and seek to prepare the student for functional analysis. Others concentrate on the visual geometric side of the subject—conic sections and quadric surfaces. Most of the recent texts try to accommodate students of engineering, economics, or biology by stressing explicit techniques to solve systems of linear equations.

Peter's *Linear Algebra* is entirely different. It requires as preparation an introductory first-year course on the subject and presents a startling variety of lesser-known mathematical results and methods with many different physical applications. It could not be widely adopted as a textbook, for few graduate schools offer a second-year course in linear algebra. But it's a delightful source of reading pleasure and a handy toolbox for a researcher on the mathematical frontier.

The general thrust and flavor of the book express Peter's pure mathematical side rather than his computational side. This is a corrective to a regrettable trend in teaching linear algebra. In his introduction, he wrote:

> During the past five decades there has been an unprecedented outburst of new ideas in linear numerical analysis, about how to solve linear equations, carry out least square procedures, tackle systems of linear inequalities, and find eigenvalues of matrices. This outburst came in response to the opportunity created by the availability of ever faster computers with ever larger memories. Thus linear algebra was thrust center stage in numerical mathematics. This had a profound effect, partly good, partly bad, on how the subject is taught today. The presentation of new numerical methods brought fresh and exciting material, as well as realistic new applications, to the classroom. Many students,

Books by Peter Lax

after all, are in a linear algebra class only for the applications. On the other hand, bringing applications and algorithms to the foreground has obscured the structure of linear algebra—a trend I deplore. It does students a great disservice to exclude them from the paradise created by Emmy Noether and Emil Artin. One of the aims of this book is to redress this imbalance. My second aim in writing this book is to present a rich selection of analytical results and some of their applications: matrix inequalities, estimates for eigenvalues and determinants, and so on. This beautiful aspect of linear algebra, so useful for working analysts and physicists, is often neglected in texts. I strove to choose proofs that are revealing, elegant, and short. When there are two different ways of viewing a problem, I like to present both.

In presenting the abstract theory of linear spaces and linear transformations, he says, "I avoid elimination of the unknowns one by one, but use the linear structure; I particularly employ quotient spaces as a counting device." Here he is taking a cue from Noether and Artin's paradise.

Chapter 11 presents a welcome bonus, "Kinematics and Dynamics". "I included it to make up for the unfortunate disappearance of mechanics from the curriculum and to show how matrices give an elegant description of motion in space. Angular velocity of a rigid body and divergence and curl of a vector field all appear naturally.

> # SCATTERING THEORY FOR AUTOMORPHIC FUNCTIONS
>
> BY
>
> PETER D. LAX AND
> RALPH S. PHILLIPS
>
> PRINCETON UNIVERSITY PRESS
> AND
> UNIVERSITY OF TOKYO PRESS
>
> PRINCETON, NEW JERSEY
> 1976

The monotonic dependence of eigenvalues of symmetric matrices is used to show that the natural frequencies of a vibrating system increase if the system is stiffened and the masses are decreased."

The question of how to solve systems of linear equations is kept waiting until the last chapter. When it comes, it doesn't review the standard direct algebraic algorithms, but instead presents iterative methods and investigates their rate of convergence. In this final chapter there is a clear, direct discussion of round-off error and this historical comment: "It is instructive to recall that in the 1940s linear algebra was dead as a subject for research. It was ready to be entombed in textbooks. Yet only a few years later, in response to the opportunities created by the availability of high-speed computers, very fast algorithms were found for the standard matrix operations that astounded those who thought there were no surprises left in this subject."

A review by Mark A. Kon appeared in the *American Mathematical Monthly* in November 2001, pages 883–887: "A graduate text in linear algebra has been quite uncommon, at least in recent years; so there has been very little competition for this niche. This book proves that there are many topics to make such an effort worthwhile, and the text indeed fills a gap in the graduate curriculum. All in all, because of its directed and original approach and the overview it brings, the book is recommended for the teacher and researcher as well as for graduate students. In fact, I think that it has a place on every mathematician's bookshelf. The proofs are direct, novel, and elegant, and the presentation inspires one to rethink material that has sometimes become too routine."

And there are two Amazon reader reviews: "If you really strive to understand what each sentence really means, it's not all that difficult...at times I spent hours reading one page." And finally, "I used to hate this book but over time it has become my favorite."

In 2002 Peter published *Functional Analysis*. This is one of Peter's specialties. The Hungarian school, especially Frigyes Riesz, played a major role in creating the subject, and the book of that title by Riesz and Nagy was one of the delights of graduate school in my day, fifty years ago.

Functional analysis is what you get by looking at problems of analysis—especially problems of differential equations, both ordinary and partial—with the tools and viewpoint of vector analysis or vector spaces. The set of continuous functions of one variable, for example, is a "linear space". Just as you can add two vectors in the plane, and you can multiply any vector by a scalar (a real number) and get another vector, so you can add a continuous function to another continuous function or multiply it by a real number and get a continuous function. This simple remark has as a consequence that you can think geometrically with respect to the space of continiuous functions. You can think of direction and angle, of rotation and reflection in that "abstract" space! A differential equation problem becomes a geometric problem in a function space.

Peter's book on functional analysis diverges sharply from purist texts motivated by abstract set theory. It "intersperses chapters on theory with chapters on applications, so that cold abstractions are made flesh and blood." It provides "a very rich fare of mathematical problems that can be clarified and solved from the functional analytic point of view." For example, after Chapter 6, "Hilbert Space", comes Chapter 7, with the Radon-Nikodym theorem and three different Hilbert-space methods of solving the Dirichlet problem. After Chapter 10, "Weak Convergence", comes Chapter 11, with sections on divergence of Fourier series, approximate quadrature, and solutions of partial differential equations.

A special feature of this book is the historical notes about creators of the subject. "Like most mathematicians, I am no historian. Yet I have included historical remarks in some of the chapters, mainly where I had some firsthand knowledge, or where conventional history has been blatantly silent concerning the tragic fate of many of the founding fathers of functional analysis during the European horrors of the 1930s and 1940s."

Here are two examples.

From page 244: "Julius Schauder (1899–1943) was the most brilliant of the Polish mathematicians of his time. Schauder bases, the Schauder fixed point theorem, the Leray-Schauder degree of a mapping, as well as many fundamental results in the theory of elliptic and hyperbolic partial differential equations, are his creation. Being Jewish, he was killed during the Nazi occupation of Poland. Such things were so routine, nobody knows when or where."

From page 172: "Stefan Banach (1892–1945), a Polish mathematician, was one of the founding fathers of functional analysis. Banach spaces are named in recognition of his numerous and deep contributions, and for having written the first monograph on the subject (1932). He was the inspiration of the brilliant Polish school of functional analysis. During the Second World War Banach was one of a group of people whose bodies were used by the Nazi occupiers of Poland to breed lice, in an attempt to extract antityphoid serum. He died shortly after the conclusion of the war. The Nazi attitude toward Poles was epitomized by the following story. After the conquest of France in 1941, when Hitler ruled most of Europe, a leading German mathematician, a member of the Nazi party, called on Elie Cartan, the dean of French mathematicians, to discuss the organization of mathematical life in the new European order. Cartan wanted to know how the Polish mathematicians would fit in. "Oh," the German replied, "the Fuhrer has declared the Poles to be subhuman."

Steven G. Krantz reviewed this book in *SIAM Review* **47** (2) (June 2005), 383–385. "The student learning from this book will garner a real education—not just in basic functional analysis *but in analysis as it is actually practiced*. This book has real power and authority just because it is written by one of the masters who helped to develop these ideas over the past 60 years."

An anonymous reader wrote on the Amazon Books webpage, "Peter D. Lax is one of the great mathematicians of the second half of the 20th century. Buy this book if only to have it on your bookshelf. At the end of a long day of suffering fools and putting up with the latest insanities of university administrators, leaf through Lax's book to recover. This is mathematics at its best. What better use can we put mathematics to than to help us regain our sanity in a world that is going mad."

Peter published two books with Ralph Phillips: *Scattering Theory* in 1967 and *Scattering Theory for Automorphic Functions* in 1976. The latter is part of Princeton University Press's series Annals of Mathematics. The Lax-Phillips scattering theory is applied to non-Euclidean geometry, resulting in a powerful new tool for analytic number theory. I comment on this material in Chapter 10.

In 2005 Peter published his *Selected Papers, Volumes I* and *II*. I reviewed these books for the *Bulletin of the American Mathematical Society* in 2006 (vol. 43, pp. 605–608). I quote from that review here as a brief summary of some of Peter's most important research and as a preliminary to the more extended discussions in Chapter 10.

> Peter Lax's mathematical work is a harmonious combination of prewar Budapest and postwar New York. By Budapest, I mean his elegant geometric, functional-analytic style of attacking hard, concrete problems. By New York or NYU, I mean his lifelong interest in physics and in practical computation....
>
> ... [He] wrote a thesis under Courant's famous student, Kurt Otto Friedrichs,... on partial differential equations, which would be Peter's lifelong mathematical milieu.... Modern physics is relativity and quantum mechanics—the Einstein equations and the Schrödinger equation.
>
> Classical physics is continuum mechanics—mainly fluid dynamics, which includes the atmosphere (meteorology) and the

Paul Garabedian (1927-2010)

oceans (oceanography)—the Navier-Stokes equation. All particular, special examples of PDE's! But what about Newtonian mechanics? Well, that's an ODE [ordinary differential equation], one of those special PDE's with only one independent variable. But the variational approach turns it into the Hamilton-Jacobi partial differential equations. Does that mean PDE equals physics? No, not quite.

Complex analysis is all about the Cauchy-Riemann equations. Differential geometry is all about the Weingarten and the Codazzi-Mainardi equations. Calculus of variations is all about the Euler-Lagrange equations.

What about topology? Well, the hottest thing there right now seems to be the Hamilton-Perelman proof of the Poincaré-Thurston conjectures on 3-manifolds. That turns out to be all about a particular nonlinear parabolic equation for "Ricci flows".

What about algebra? Well, PDE's with constant coefficients are simply polynomial equations with differential operators as their variables. Consequently, the essence of that simplest part of PDE is the connection between algebraic conditions on the polynomial and analytic properties of the solutions; thus the classification, originally of second-order and, in modern times, of general constant coefficient PDE's into "elliptic", "parabolic", "hyperbolic", "Petrowsky-correct" and so on types. Here, rather than contribute solutions to algebra, PDE contributes problems, new and challenging ones.

[In 1960 Peter published a conjecture on] hyperbolicity: i.e., how to represent a certain polynomial all of whose roots are real.... Peter's conjecture was recently proved by A. S. Lewis et al. The proof hinged on an observation of Helton and Vinnikov, based on a deep result of Vinnikov on self-adjoint determinental representation of real plane curves which Vinnikov published in 1993. Thus Peter Lax's specialty in partial differential equations opened up all of mathematics and physics to his enjoyment.

These two books were edited by Andy Majda and Peter Sarnak. The selected articles are divided into nine parts: Partial Differential Equations, Difference Approximations to PDE, Hyperbolic Systems of Conservation Laws, Integrable Systems, Scattering Theory in Euclidean Space, Scattering Theory for Automorphic Functions, Functional Analysis, Analysis, and Algebra. After each part there are a few pages of commentary, some by Peter, some by others. All together, 58 papers have been selected, from a list of 158 publications over almost sixty years (1944 to 2003). Item 34 in the publication list is the important paper, co-authored with Burt Wendroff, "Difference schemes for hyperbolic equations with high order of accuracy". Somehow, in the table of contents, this paper is mistakenly listed as co-authored with "W. Burton", which is not much of an alias for "Burt Wendroff". It is striking that of the 58 papers in this Selecta, only 13 papers deal with difference approximations. Of the 158 papers in the CV, around 30 seem to be about finite difference approximations.

After this book was published, with its complete list of Peter's publications, he published three more books, which of course are not included in that list. *Hyperbolic Partial Differential Equations* originated in lectures at Stanford University in 1963. Lax updated and enlarged them, and the book was published jointly by the Courant Institute and the American Mathematical Society in 2006. I used it in a seminar at the University of New Mexico. It begins with his own high-level axiomatic treatment of hyperbolicity. Instead of giving dogmatically the definition of a hyperbolic equation, he starts qualitatively, with an axiom of finite speed of signal propagation. From this qualitative axiom, the algebraic condition of "hyperbolic" is derived.

Peter wrote: "The first seven chapters of this book deal with basic theory: the relation of hyperbolicity to the finite propagation of signals, the concept and role of characteristic surfaces and rays, energy, and energy inequalities. The structure of solutions of equations with constant coefficients is explored with the help of the Fourier and Radon transforms. The existence of solutions of equations with variable coefficients with prescribed initial values is proved using energy inequalities. The propagation of singularities is studied with the help of progressing waves." He doesn't even bother to mention the elegant chapter where pseudo-differential operators are used to study linear equations with variable coefficients. These operators were introduced by Calderón and Zygmund as singular integral operators. They reached a more abstract, systematic presentation in work of Kohn and Nirenberg. Compared to other presentations, Peter Lax's version is a marvel of conciseness and concreteness. It is all done very explicitly and almost elementarily, just what is needed to get the job done.

Chapter 8 is about finite difference approximation. Chapter 9 is on scattering theory. Peter said to Phil Colella in their interview that this chapter is better than what's in his book on scattering theory. A few years after doing the research, he was able to do a better exposition. Chapter 10 departs from linearity to present hyperbolic systems of conservation laws.

Appendix D is about the mixed initial-boundary value problem. There, and in his interview with Phil Colella, he gave credit to Heinz-Otto Kreiss and Reiko Sakamoto, but in fact my 1962 thesis was the starting point from which both of these two impenetrable contributions generalize.

Complex Proofs of Real Theorems, with Lawrence Zalcman, was published in 2012 as volume 58 in the American Mathematical Society University Lecture Series. The title is a double entendre, recalling a bon mot of Paul Painlevé (often misattributed to Jacques Hadamard): "Between two truths of the real domain, the easiest and shortest path quite often passes through the complex domain." Lax and Zalcman present nineteen theorems whose statements involve only real numbers, but whose proofs pass through the complex plane. There is a "Coda", with summaries of two impressive applications to fluid dynamics and statistical mechanics.

In case you haven't seen it, here's an event, paradoxical on the real line, that's easily explained in the complex plane. The function $f(x) = 1/(1+x^2)$ is the sum of a geometric series whose nth term is $(-x^2)^n$. For all real x, this function is perfectly smooth and well behaved, yet the series blows up for x greater than or equal to 1 in absolute value. What makes it blow up? To answer this question, extend the function f and its geometric series from real x to complex z. Then you notice the singularity at $z = i$. For $z = i$, $f(z)$ does not exist—it is infinite. That's why the power series can't converge on or outside the unit circle for complex numbers with modulus greater than or equal to 1.

The high point of this text is a surprisingly easy proof (only six pages long) due to Donald Newman, of the famous Prime Number Theorem. In the Coda there is described an application to shock waves, one of Peter Lax's main long-term specialties. It concerns the design of airplane wings and was done by Paul Garabedian, Peter's long-time colleague at the Courant Institute. Garabedian specialized in daring and unconventional applications of complex analysis. With a collaborator, David Korn, he was able to calculate supercritical wing sections free of shocks at specified speed and angle of attack. The key step in this tour de force was solving the partial differential equations of two-dimensional inviscid gas dynamics by analytic continuation into the domain of two independent complex characteristic coordinates.

In 2012 Peter published, with his collaborator Maria Terrell, volume 1 of *Calculus with Applications*, the new improved version of his calculus text (functions of one variable). Volume 2, on functions of several variables, is forthcoming.

CHAPTER 9

Pure AND applied, not VERSUS applied

This chapter is preliminary to Chapter 10, where I explain and describe some of Lax's major mathematical achievements. The present chapter starts with a discussion of "pure mathematics" versus "applied mathematics". Then I discuss the interaction between high-speed computing and mathematical research. There follow five brief tutorials. The first explains how Peter's kind of mathematics comes out of familiar physics. Then I offer introductions to four themes of pure and applied mathematics: partial derivatives, phase space, function space, and nonlinearity. These tutorials are meant to prepare the mathematically unsophisticated reader for Chapter 10, where, in nontechnical language for a nonprofessional audience, Peter's major mathematical contributions are presented.

The division of mathematics between "pure" and "applied" is a misleading cliché. In fact, the phrase "pure mathematics" is rarely used nowadays. You call yourself "mathematician", tout court (or "algebraist", "geometer", "analyst", "number theorist", "probabilist"). The term "applied mathematics" is also a bit outdated or old-fashioned. Nowadays people say they do "mathematical modeling" or "scientific computation".

Marshall Stone once gave this definition of "modern mathematics", which could serve as some people's notion of pure mathematics: "The study of abstract systems, each one of which is built of specified abstract elements and structured by the presence of arbitrary but unambiguously specified relations among them."

Even more off-putting are Bertrand Russell's pronouncements that "in 1854 George Boole discovered pure mathematics," and that "mathematics may be defined as the subject in which we never know what we are talking about, nor whether what we are saying is true" (*Mysticism and Logic*, Chapter 4, 1917).

Instead of "pure" and "applied" it might be better to call it "non-applied" and "applied". "Matrices" were invented by Arthur Cayley, an English algebraist whose mathematics was as pure as the Virgin of Guadalupe. But now Cayley's matrix algebra has become the central core of physics and applied mathematics. Michael Atiyah and Alain Connes are high-prestige mathematicians, never called "applied mathematicians". But both of them have become entangled with elementary particle physics. Algebraic number theory was always the purest of the pure, until a few years ago. Today, an esoteric multiplication based on elliptic curves is being sold to bankers and CIA spooks to conceal financial and military secrets. Once pure, now dirty!

The sociologist Bernard Zarca reports that French probabilists resist classifying themselves as either pure or applied; they prefer to call themselves just mathematicians. That certainly would be Peter Lax's preference. He has said, "I never felt in conflict, or even contrast, between pure and applied mathematics. The thinking needed in either was quite similar."

In "Mathematics and its applications", a talk delivered at Oberwolfach in 1984 and published in the *Mathematical Intelligencer* in 1986, Peter said: "The separation of mathematics into pure and applied is a recent—and transitory—phenomenon For Poincaré, Hamilton, Maxwell, Stokes, Kelvin, Rayleigh, Boole, Gauss, Riemann, Klein, Hilbert, Gibbs there was no such separation. Gauss, Prince of Mathematicians, was also Prince of Computing; for example, his Nachlass shows that he made use of the Fast Fourier transform. The bold proposal to cut the lifeline between mathematics and the physical world was put forth only in the twentieth century, mainly by the Bourbaki group. Besides being wrong-headed, this raises profound philosophical problems about value judgments in mathematics. The question 'What is good mathematics?' becomes a matter of a priori aesthetic judgment, and mathematics becomes an art form. There is, of course, some truth in this, but it seems to me that as art, mathematics resembles most closely painting. In both there is a tension between two tasks; in painting, to present the shapes and colors of the visible world, and also to make pleasing patterns on a two-dimensional canvas; in mathematics, to study the laws of nature, and also to spin beautiful deductive patterns. The most successful creations are those where the tension between these tendencies is greatest; the least satisfactory are those works where one aspect predominates, as in genre painting or pure abstraction."

These remarks of Peter's reveal his deep interest in the art of painting. In his twenties he studied for seven years with a woman he calls wonderful. Her name was Anna Lesznai. She was Hungarian. Peter's parents were acquainted with her. She exhibited as early as 1911 in a style that combined Art Nouveau with Hungarian peasant influences. She was part of the famous Budapest Sunday Circle, created by the film critic Béla Balász, and including the Marxist philosopher Georg Lukacs. After the crushing of the Budapest Soviet revolution in 1918, she fled to her parents' country estate and then to Vienna, where she lived until 1939. Then with her husband she fled to New York. Peter still has stacks of nude studies that he created as an art student.

Peter Lax's work merges two great traditions of modern mathematics—from Budapest and Göttingen, a special style of functional analysis and from Los Alamos, a certain approach to machine computation and numerical analysis. His contributions are part of a long tradition where interaction between mathematics and physics is at the core. Physics offers challenging problems that require intuition to solve. Mathematics can reveal to physics its deep inner structures, and mathematical rigor can provide a solid foundation for physical knowledge.

"Applied mathematics" is sometimes misinterpreted to mean ignoring rigor, resorting to guesswork and unjustified simplifications in order to get a numerical result. Peter was never that kind of applied mathematician, but rather a pure mathematician who rigorously solved mathematical problems that have origins and applications in physics.

It would take us very far afield to try to straighten out this whole tangle; all we need to do in this book is to see what Peter Lax actually does or did and to relate it to the milieu of his time.

While in Oslo to receive the 2005 Abel Prize, Peter Lax was interviewed by the Danish mathematician Martin Raussen and the Norwegian Christian Skau at that city's Hotel Continental (*Notices of the AMS*, February 2006).

Anna Lesznai, artist and poet with whom Peter studied painting

Raussen and Skau asked, "Are there examples where your theory of nonlinear partial differential equations, especially your explanation of how discontinuities propagate, have had commercial interests? In particular, concerning oil exploration, so important for Norway!"

Lax answered, "Yes, oil exploration uses signals generated by detonations that are propagated through the earth and through the oil reservoir and are recorded at distant stations. It's a so-called inverse problem. The direct problem is calculating how signals propagate, when you know the distribution of the densities of materials and the associated waves' speeds. The inverse problem is the opposite—when you know how signals propagate, and you want to deduce the distribution of the materials. The signals are discontinuities, so you need the theory of propagation of discontinuities. It's somewhat similar to the medical imaging problem, which is also an inverse problem. There the signals go, not through the earth, but through the human body. You have to understand the direct problem very well before you can tackle the inverse problem."

They also asked about some clashes between the pure and the applied. "Occasionally one can hear within the mathematical community statements that the theory of nonlinear partial differential equations, though profound and often very important for applications, is fraught with ugly theorems and awkward arguments. In pure mathematics, on the other hand, beauty and aesthetics rule. The English mathematician G. H. Hardy is an extreme example of such an attitude, but it can be encountered also today. How do you respond to this? Does it make you angry?"

Peter answered, "I don't get angry very easily. I got angry once at a dean we had, terrible son of a bitch, destructive liar. I got very angry at the mob that occupied the Courant Institute and tried to burn down our computer. Scientific disagreements do not arouse my anger. But I think this opinion about ugliness and beauty is wrong. Paul Halmos once claimed that applied mathematics was, if not bad mathematics, at least ugly mathematics. I could answer by pointing to the citations of the Abel Committee dwelling on the elegance of my works."

In his interview with Colella, he spoke about mathematical elegance, this time with a Hungarian flavor. "There was a tradition in Hungary to look for the simplest proof. You may be familiar with Erdős's concept of The Book. That's The Book kept by the Lord. In it are all theorems and their best proofs. The highest praise Erdős had for a proof was that it was 'out of The Book.' One can overdo that. Shortly after I got my Ph.D., I learned about the Hahn-Banach theorem, and I saw that it could be used to prove the existence of Green's function. It's a very simple argument—I believe it's the simplest—so it's out of The Book. And I have a proof of Brouwer's Fixed Point Theorem using calculus, just change of variables. It is probably the simplest proof, and again, it's out of The Book. All this is part of the Hungarian tradition. But one must not overdo it."

Colella asked, "One of the unspoken rifts between computational mathematicians or numerical mathematicians and the more traditional pure mathematicians is that in computing there are, if not no proofs, precious few. And that makes us a breed apart, from the point of view of traditional mathematics. Is there some hope that maybe that rift will be closed, as pure mathematicians use computers to manage their complexity?"

Peter answered with an anecdote about the Swedish numerical analyst Heinz Kreiss. "Kreiss sent his first paper to *Mathematica*, which was a very distinguished and very theoretical journal, and the editor sent it to me asking, 'Is this the kind of thing that we should be publishing?' And I said yes." [*laughter*]

Lax went on, "There is really a different point of view. Take three-dimensional or even two-dimension compressible-flow calculations in air, or even the very simple problem of diffraction by a wedge, which produces a double Mach refraction. I don't think we will ever be able to prove that the computed solution is within a prescribed epsilon of the true flow. That's hopeless. And if there were such a proof it would be unenlightening. It's far more convincing that a calculation carried out by an entirely different numerical method gives a very similar result. In that respect, the computational mathematician does indeed differ from the theoretical mathematician. It's a different style. I don't think the two need to be reconciled. Von Neumann, who was certainly a marvelous theoretical mathematician, also enjoyed doing calculations. It never entered his head to prove rigorously that they were within epsilon of the solution. He made some calculations for designing the implosion employed in the plutonium bomb. They were not used by the physicists because they were so novel, they were unacceptable."

Lax even insists that the stability theory of finite difference approximations is MORE difficult, more subtle, than the more prestigious theory of differential equations.

As to the acceptability of computer proofs, Peter recalled that "the computer proof of the Four Color Theorem was criticized because it does not give insight into why the result is true, but such criticism can also be leveled against proofs that don't use computing. It is amusing that when logicians were looking for a rigorous definition of a mathematical proof, they came up with Turing's definition of calculation carried out by a Turing machine. [*laughter*] So calculation carried out by an imaginary computer was the epitome of mathematical exactitude! But that left open the right to argue against proofs employing a real computer. However, there is a different problem that really has to be addressed, for both very complicated conventional proofs and computer-assisted proofs: how to make them more transparent. On the question of reliability, computer science can be helpful with program verification, which is a branch of computer science."

Colella: But then you have this potential for an infinite regress: how do you verify the verification program?

Lax: Sure, sure. Well, what would you say about the classification of simple groups? At the Phoenix meeting in January I listened to an interesting lecture by Michael Aschbacher from Caltech. He is one of the leaders in this work. He said that there were some gaps in it, but now they have been closed. You could ask the same question: how do we know that they have really been closed? That's not unique to computer-assisted proofs. But they have to be tackled by different methods. To certify a proof as correct, people really have to understand it.

Peter Lax's "applied mathematics" doesn't involve creating new models of physical reality. In the fields where he has worked—fluid dynamics, electromagnetism, acoustics—the "continuum" model created by Isaac Newton and Leonhard Euler still is effective. The basic equations like the Euler or Navier-Stokes equations have been accepted for centuries. So it's not called mathematical physics; it's called applied mathematics.

His works on these models, on the equations of smooth flow and their modifications to describe shock waves, are pure analysis, beautiful works of "pure mathematics". That is to say, they are about the mathematical equations, which are formulas of differential and integral calculus. He establishes existence, uniqueness, and regularity or irregularity of their solutions, and gives representations—"explicit formulas"—for the solutions. There is strict rigor, even though many details of complete proofs are omitted, left for the reader to fill in.

Yet these same results are applied mathematics! The problems arise from physics and are useful in physical and engineering applications. A shock wave, for example, speaking mathematically, is merely a function with a certain kind of discontinuity. But a "shock wave" also is a clap of thunder, a sonic boom, an explosion or an implosion, nuclear or nonnuclear.

Nearly all of Peter's work has been on physical problems that are expressed mathematically as "partial differential equations of hyperbolic type". These equations govern wave propagation, including sound waves and light waves. This part of classical physics has been studied mathematically for centuries. The basic ideas and formulas were known already in the 1700s. Yet many central problems remain unsolved. And wave propagation reappears in quantum mechanics and general relativity.

In Chapter 10 we look in more detail at four fields where Peter has been a transformative leader: finite difference approximations for partial differential equations; solitons as solutions of integrable nonlinear PDE's; shock waves and nonlinear conservation laws; and the scattering theory of classical wave propagation, including acoustics, light, and electromagnetism. These major works of Peter's are both pure and applied. They combine his virtuosity with computers and his virtuosity in functional analysis. It becomes clear why, in his experience, "both pure and applied used the same kind of reasoning." They are applied because the problems originate in physics and engineering and the results are used in physics and engineering, but they are pure because the mathematical work is perfectly rigorous and uses abstract theories of functional analysis.

While in his own work rigor and applicability are inseparable, he is well aware that much of applied mathematics does manage without logical rigor. He wrote, "Having described some of the achievements of applied mathematics, I would like to discuss briefly its methods. Some of them are organic parts of pure mathematics: rigorous proofs of precisely stated theorems. But for the greatest part the applied mathematician must rely on other weapons: special solutions, asymptotic descriptions, simplified equations, experimentation both in the laboratory and on the computer. Out of these emerges a physical intuition, which serves as a guide to research. Since different people have different intuitions, there is a great deal of controversy among applied mathematicians. It is a pity that these debates so often become acrimonious, shedding more heat than light" ("The flowering of applied mathematics in America", p. 451).

Computing

As a general rule, if you want more than qualitative information about a PDE, even a linear one with variable coefficients, and especially a nonlinear one, your only hope is to approximate it. Of course, the point is to approximate it with equations you can solve by partial *difference* equations whose solutions can be "written down" recursively. One major piece of Lax's mathematical research has been inventing finite difference approximations to solutions of partial differential equations. This is "numerical analysis", an indispensible branch of applied mathematics. The problems in fluid dynamics where Peter found explicit solutions are very special and simple compared to real engineering problems such as fission and fusion bombs or airfoils for modern aircraft. Hard-core engineering models have to be approximated; they can't be solved "explicitly". Engineers demand approximations that are accurate, reliable, and not too slow or expensive. In this kind of mathematics, rigorous proof is rare. The algorithms embodied in a computer code may have been rigorously justified only in a special model example. However, they will then be used on massive numerical problems. Even in this area, Peter was sometimes able to obtain rigorous proofs. But sometimes he enjoyed relinquishing the constraint of rigor for the sake of useful results.

For "customers" in science and industry, the absence of rigorous proof is rarely a concern. Practical results are the thing. "Does this algorithm work?" "Well, yes, it seems to work as far as I can tell." That might be good enough for the customer.

Then the mathematician may go back to her office and struggle for the thatrigorous proof is still out there, waiting to be discovered.

To many practitioners of pure mathematics, algorithmic or numerical mathematics, sometimes referred to as scientific computation, hardly counts as mathematics at all. A high-prestige mathematical prize is seldom awarded for such work. On the other hand, the numerical analysts have established their own prizes. Useful numerical schemes are highly appreciated in the "real world" of industry, both peaceful and military. Numerical analysts are found in departments of engineering, as well as in departments of mathematics. Some universities have separate departments of applied mathematics, where numerical analysts are welcome.

The study of finite difference approximations to solutions of partial differential equations is a long-standing major area of applied mathematics. But Peter Lax is virtually alone in insisting that it is also a major field for pure analysis—indeed, more difficult than the traditional existence and uniqueness theory. His calls for participation by other pure analysts have not received much response over the years. He stands almost alone as a master of both the computational and the theoretical sides of the subject.

Peter Lax said, "High speed computers' impact on mathematics, both applied and pure, is comparable to the role of telescopes in astronomy and microscopes in biology." The logical construction of computers and their operating systems is by nature mathematical. But computers also serve as laboratories for mathematicians. New mathematical relations can be discovered, and our hypotheses and assumptions can be disproved or found more likely by computations. Lax mentioned George David Birkhoff's lifetime effort to prove the ergodic hypothesis before computers were available. If Birkhoff had been able to test the hypothesis on a computer, he would have found examples where it isn't true. Problems of modern technology such as the simulation of airplanes, oil rigs, or the weather require not only very powerful computers but also the development of new and better mathematical algorithms for their solution. The development of high-speed computers (hardware) and the development of new numerical techniques (software) contributed equally to the total performance we observe in simulations. Peter Lax has made penetrating contributions to the development of many of these new mathematical methods.

Paul Garabedian was another mathematician at the Courant Institute besides Peter who had feet firmly planted in both the pure and applied sides. Garabedian first won fame for contributions to the Bieberbach conjecture on conformal mapping, and then won fame for designing airfoils free of shock waves. Garabedian told me that he had solved the energy crisis by calculating how to maintain stability in a contained fusion reaction. In that same conversation, he confided that not only had he once been a child prodigy, he was still a child prodigy "but nobody knows that."

In Peter's estimation, "Paul was one of the most original applied mathematicians that I know, also a distinguished pure mathematician. He was among the first to solve the reentry problem. When intercontinental missiles were first conceived it wasn't clear that they could reenter the atmosphere without being destroyed, as the *Challenger* was. It would have been better for the world if this problem had no solution. In the late 1960s Paul turned to the design of shockless airfoils; that was an enormously original piece of work. He complexified the space coordinates and turned the elliptic equations to hyperbolic ones. No aeronautical engineer would believe in such a thing, and so they refused to test his designs. He had to go to Canada, where he found an aeronautical engineer who was more sophisticated mathematically and believed Paul, built the airfoil Paul designed, and tested it. Indeed it was shockless. Then, of course, everybody jumped on it. He also had other good ideas, a very neat analysis of over-relaxation. The last twenty years Paul has been working on fusion."

Garabedian received the George David Birkhoff Prize of the AMS and SIAM in 1983 and the Theodore von Kármán Prize of SIAM in 1989.

In an interview with Phil Colella, Peter said, "In the old days, to get numerical results you had to make enormously drastic simplifications to do your computations by hand or by simple computing machines. That didn't appeal to most mathematicians. Today you're in an entirely different situation. You don't have to put the problem on a Procrustean bed and mutilate it before you attack it numerically. And that has attracted a much larger group of people to numerical problems of applications—you can really use the full theory. This invigorated the subject of linear algebra, which as a research subject died in the 1920s. Suddenly the actual algorithms for carrying out these operations became important. There were lots of surprises, like fast matrix multiplication. In the new edition of my linear algebra book I add a chapter on the numerical calculation of the eigenvalues of symmetric matrices.

"You know, it's a truism that due to increased speed of computers, a problem that took a month forty years ago can be done in minutes or seconds today. Most of the speed-up is attributed by the general public to increased speed of computers. But actually, only half of the speed-up is due to this increased speed. The other half is due to clever algorithms, and it takes mathematicians to invent clever algorithms. So it is very important to get mathematicians involved, and they are involved now. Computational science, computational mathematics, has a very bright future, and I warmly recommend that young people go into that field."

In the mathematical scene, Peter Lax is a rarity, equally comfortable in both worlds. Even at Courant, the faculty can be divided into the computationally oriented and the theoretically oriented. Peter's mentor, Kurt Friedrichs, and his closest mathematical friend, Louis Nirenberg, while certainly interested in the applied side of PDE, never got into numerical work. Friedrichs was co-author of the famous Courant-Friedrichs-Lewy paper of 1928. They used finite difference equations for establishing results about partial differential equations even though high-speed computers did not yet exist. When such computers did arrive, Friedrichs was not tempted to employ them.

The two-sided nature of Peter's mathematical personality can be seen by looking over the list of his fifty-five Ph.D. students. As far as I can tell, every single student can be clearly characterized as either "pure" or "computational". Several of the most successful devote themselves mainly to numerical computing, along with some modeling. They don't focus on mathematical rigor or publishing in high-prestige pure math journals. Others (myself included) taught standard math courses in math departments and worked on rigorous proofs about existence, uniqueness, etc., of solutions of partial differential equations.

I wrote two papers on finite difference approximations, but I never tried to implement either of them in any particular example. That would have meant writing "code", which I had no interest in learning. It would have meant finding somebody with a real problem who could use the theory I had developed—probably an engineer or physicist, who would speak a different vocabulary. My attitude was, "I did the mathematical part; it's up to somebody else to bring it down to the practical level." Of course, nobody felt motivated to bother. My attitude was acceptable, even normal and typical, among professors of mathematics. Applications are fine, but not something we need to struggle with.

It is thought-provoking that while in Lax's work the pure and the applied are inseparable, his students identify strongly with one or the other. I think that Peter discerned a student's natural inclinations and suggested problems in accord with that student's inclinations. Stanley Osher is a remarkable example. Peter diverted him from pure to applied with brilliant results.

Numerical analysts may not be aware that three quarters of Peter's mathematical output is theoretical, not numerical. Conversely, theoretically minded readers of Peter's two-volume *Selected Papers* may not notice that a part of Peter's work is algorithmic, proposing computational methods based on plausibility and test results. He has worked with collaborators and programmers, who helped in the detail work, writing code, and running trials. He urged Alex Chorin not to spend time writing code. But his conversations with Colella show intense interest in faster schemes and more useful outputs.

"The role of numerical mathematics in computational science is very great, because mathematicians have the most fertile imagination in inventing new methods. I think they are better than physicists or chemists—I hope that won't bring the wrath of physicists and chemists on my head! Many wonderful computing methods were invented by physicists, such as Monte Carlo and the so-called Metropolis algorithm, which in fact is due to Marshall Rosenbluth.

"Computation is a discipline that sits between a lot of other more traditional disciplines. Where should it be inside a university? The computer science departments, most of them, are not a natural home for scientific computation. The one at Courant used to be an exception, but that's in the past. Computer scientists don't compute, in the sense of scientific computation. Mathematics is the natural home. It would be worthwhile to look through the leading universities and see how they handle computing.

"It's quite clear that the relation between mathematics and computing will be with us forever, because new problems will come up, new machines will come up, new computing methods will be invented. I think it will always be on the leading edge of science. In a talk to the American Philosophical Society I made this point: it's a measure of the importance of computing that if you look at Nobel Prizes in physics, more and more of them depended on computing. The first one was for CAT scan. It honored Alan Cormack for his numerical contribution. Then in 1986 Herbert Hauptman and Jerome Karle were awarded a Nobel Prize for X-ray crystallography, in which the key idea was a numerical implementation of the Toeplitz/Caratheodory characterization of the Fourier transform of positive distribution. Walter Kohn and John Pople shared a Nobel Prize in chemistry for a computational study of large molecules."

Partial differential equations and the real world, a beginner's primer

The world of classical physics is the world we live in: weather and climate, automobiles and airplanes, breathing and heartbeat. These familiar phenomena are studied and analyzed with the classical mathematics of Isaac Newton and James Clerk Maxwell, updated and magnified by functional analysis and electronic computers.

Readers who understand the expression "nonlinear hyperbolic partial differential equation" may wish to skip ahead to Chapter 10. For those who aren't familiar with these terms, here is a nontechnical introduction to some important phenomena: light and sound, temperature and pressure, diffusion, equilibrium and signal propagation. You don't have to remember any differential or integral calculus; I offer an elementary explanation of what is a "partial differential equation" and why it matters. The best way to do that is by a physical example.

Take a metal bar capable of conducting heat. It is wrapped in asbestos so heat can't pass through the sides, only through the ends. If you stick one end into an oven and the other into a tub of ice water, heat will travel through the bar from the hot end to the cold end. What is the temperature $T(x,t)$ at a particular point x and a particular time t? This is the problem of heat conduction.

Following Isaac Newton and Leonhard Euler in the eighteenth century, we model heat as a fluid that flows from a hotter place toward a colder one. During a time interval and across a section of the bar, any increase in the total amount of heat energy must equal the amount that has flowed in, minus the amount that has flowed out. Stated as a mathematical equation, there are four terms: the integral of the temperature T over the section of the bar, evaluated at both an earlier time and a later time, and the "flux" of T in and out of the two ends during the given time interval. These integral expressions are greatly simplified by shrinking the time interval and the length of the section of bar to zero, thereby bringing in the two "rates of change" or "partial derivatives" with respect to time t and position x. (Newton called them "fluxions" or "flowings".) These partial derivatives are denoted as "dT/dt" and "dT/dx".

The conservation equation then becomes an equation between derivatives. It turns out that the rate of change with respect to time, dT/dt, will be proportional to the *second derivative* with respect to x—the gradient of the gradient of the temperature. The resulting equation is

$$dT/dt = k d^2 T/dx^2,$$

the one-dimensional heat equation, our first example of a partial differential equation. The word "partial" reminds us that since the temperature T depends on two variables, x and t, it has two different kinds of derivatives. The numerical coefficient k characterizes the medium and is called diffusivity. It is different for steel and copper.

By writing this equation we have turned the physical problem into a mathematical problem. In order to solve it, we need to know how fast heat is being supplied by the oven, how fast it is being absorbed by the ice water, and what was the original heat distribution in the bar when the experiment started. With the heat equation plus these additional pieces of information, we have a "well-posed initial-boundary value problem". There are standard mathematical tools for solving this problem about a partial differential equation (PDE) with prescribed value at $t = 0$. There is exactly one function $T(x,t)$ that satisfies it.

If the flux of heat into the hot end of the bar and out of the cold end are kept constant and balanced, the temperature in the metal bar stabilizes. At any given point on the bar, the temperature ultimately attains a limiting value. This final

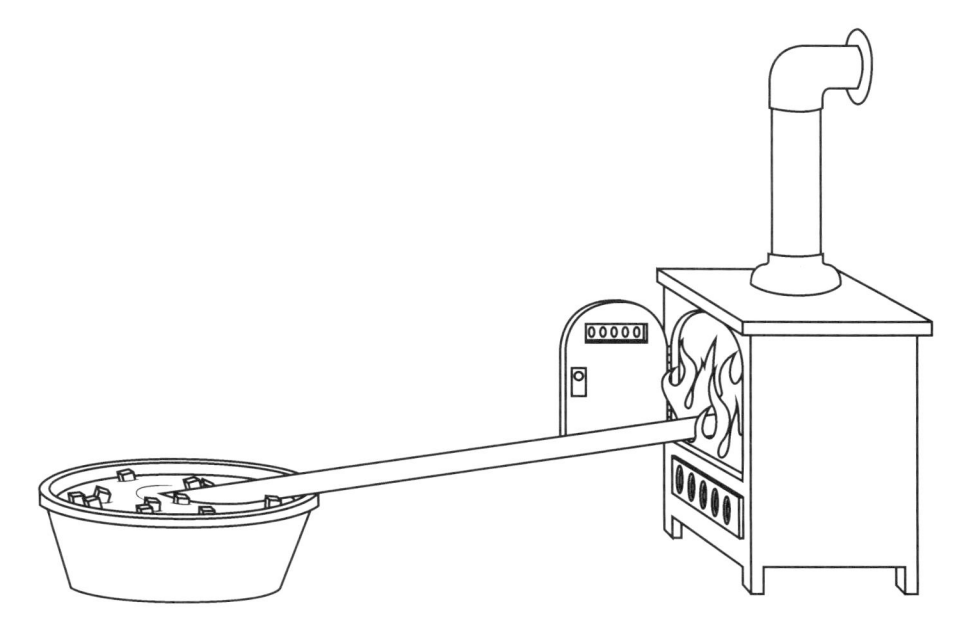

FIGURE 1. One-dimensional heat flow experiment.

"steady state" or "equilibrium" temperature is a function only of the position x and the steady temperatures at the two ends. Of course, along the bar the temperature varies from the hot end to the cold end. This is called a "boundary-value problem". It is clearly simpler than the "initial-boundary value problem", which asks for the "transient" temperature changes that take place before equilibrium is attained.

Our metal bar story has served as an example of two different types of problem. The final, equilibrium temperature distribution can be calculated as the long-time outcome of the transient distribution or it can be obtained directly from the two boundary conditions, ignoring the time history before it is attained.

A different heat conduction problem arises if the bar is very long or the time period very brief, so that the influence of the end points can be ignored. Suppose that the experiment starts with a very brief, very concentrated initial pulse of heat at a point we choose to be the "origin" of the x-axis, $x = 0$. The heat energy will diffuse to the right and left. At any fixed point x, the temperature $T(x, t)$ will rise to a maximum with respect to time t and then decrease to temperature $T = 0$. In this heat experiment, heat diffuses out through the medium from an initial concentrated impulse. A mathematically identical experiment is a drop of cream diffusing through a long, straight cup of coffee! So the heat equation is also called "the diffusion equation".

Equations like the heat equation are called "parabolic". We now turn away from diffusion and parabolic equations to consider another type of PDE, the "hyperbolic", which model signal propagation. To get a mathematical equation for signal propagation, we use a principle of "conservation of information", namely, the fact that the signals our eyes and ears pick up are not too distorted from their original source. Suppose a signal is moving without distortion to the right along the x-axis. Let $w(x)$ be the signal strength as function of position at the initial time, $t = 0$. If the signal moves to the right at some constant speed c, the state of the medium at some future time t will be given by shifting that curve or graph to the right a distance ct (because distance equals speed c times time t). The signal intensity $w(x,t)$ at a point x at time t will be equal to the original intensity w a distance ct to the left of x. So if the state of the medium at position x and time t is denoted by $f(x,t)$, and if the initial state, $f(x,0)$, was given by $w(x)$, then at time t we will have $f(x,t) = w(x - ct)$.

Taking partial derivatives, we see that the partial derivative of f with respect to x is $w'(x - ct)$ and with respect to t, it is $-cw'(x - ct)$. Putting these two equations together, we find that f satisfies this simple partial differential equation of first order, called a "transport equation":

$$df/dt = -cdf/dx.$$

If the signal travels to the left instead of to the right, we could just take c to be negative, but we prefer keeping the speed c as a positive number and describing a signal moving to the left by the formula $f(x,t) = w(x + ct)$, so that the transport equation satisfied by f is now

$$df/dt = cdf/dx.$$

These two separate cases can be written as a single equation by taking second derivatives. We get "the one-dimensional wave equation",

$$d^2 f/dt^2 = c^2 d^2 f/dx^2.$$

The letter f still stands for the amplitude or intensity of the signal, the sound wave or light wave.

This equation, like the diffusion equation, is "linear" as long as the wave speed c does not depend on the intensity f or, for the diffusion equation, as long as the diffusivity k does not depend on the temperature T. Unlike the first-order transport equations, this "second-order partial differential equation" is not uniquely determined by its initial value $f(x,0)$. We must also specify an initial velocity:

partial of f with respect to t at time $t = 0$

in order to make the motion uniquely determined. This is easy to understand by thinking of the motion of a guitar string. The sound created by *striking* a guitar string (giving it a nonzero initial velocity) is different from the sound created by *plucking* it and then releasing it (giving it a nonzero initial displacement and a zero initial velocity.)

In two or three dimensions, the mathematics becomes much more interesting. It also becomes more interesting if we do not assume that the signal speed c is a constant. If the speed of the signal is affected by the displacement from equilibrium so that c is some function of f, the "unknown" amplitude of vibration, we have a much more interesting and complex problem. This kind of equation, where the coefficients depend on the unknowns, is "nonlinear". In Chapter 10 in the section on shock waves, we solve *even the nonlinear* one-dimensional transport equation in an elementary way. But the motion of the medium may then become discontinuous! In two or three dimensions, to this day there is no general theory. To study shock waves in several dimensions we are compelled to use computer modeling and numerical approximations.

Thinking geometrically about dynamical systems: Phase space

Take the simplest possible example, a stone tossed upward with some initial velocity v. We use the letter y for its distance above the ground. The derivative of y with respect to t, denoted as y', is its speed. Now we make a daring move and locate the position y and speed y' by a point in a plane with two perpendicular axes, a vertical y axis and a horizontal y' axis. A pair of numbers y and y' represent position and speed. The history of the stone's rise and fall will be represented by a curve in the $y - y'$ plane. The time t is now a parameter for the evolution of both y and y'. Neglecting friction, the downward velocity at which it falls back to the ground will be the negative of its initial velocity upward, so the flight starts at

$$y = 0, y' = v$$

and ends, hitting the ground, at

$$y = 0, y' = -v.$$

From the starting point to the ending point in the $y - y'$ diagram, there is a curve. This trajectory in "phase space" is the history of the motion. A geometric picture, a curve, becomes available to analyze the kinematics of motion in one dimension. Phase space is a kind of abstract or fictitious geometry which permits the physicist to think geometrically about mechanical systems. The "degrees of freedom" of the mechanical system becomes the "dimension" of the phase space.

For the stone rising and falling, we can calculate the equation of its trajectory in phase space. Near the surface of the earth, gravity imparts a constant downward acceleration g, which is equal to y'', the rate of change of the speed y'. So we have a simple differential equation,

$$y'' = -g.$$

This equation says that $y''(t)$, the derivative of $y'(t)$, is constant. A function whose derivative is constant is linear, so $y'(t)$ must be a linear function. Since v is the value of the velocity $y'(t)$ at $t = 0$, we get

$$y' = -gt + v.$$

We can solve this equation for t,

$$t = (v - y')/g.$$

Integrating again, we get

$$y = -gt^2/2 + vt.$$

Now plug the expression for t into this formula for y. The time variable is eliminated, and we have an equation in $y - y'$ phase space, which can be simplified to

$$y = (-y'^2 + v^2)/2g.$$

We recognize this! This equation between y and y' is the equation of a parabola! So the trajectory in the $y - y'$ phase plane is an arc of a parabola, starting at $\{y = 0, y' = v\}$, moving downward and to the right, and ending at $\{y = 0, y' = -v\}$, with a maximum value of y at $\{y = v^2/2g, y' = 0\}$.

If we were interested in two particles simultaneously moving along an x-axis, we would need four numbers to describe this little "system" for the position and speed of the two particles. Even this tiny little problem moves us up into a four-dimensional phase space, where the time evolution of this two-particle system is represented by a curve in 4-space. But motion in one dimension has limited possibilities. We are more interested in motion in the plane. A single particle moving in the plane requires a four-dimensional phase space: two numbers to specify a position and two numbers for velocity. A single particle moving in 3-space needs a six-dimensional phase space, and a system of n particles in 3-space generates a curve in six-n-dimensional phase space. n could be nine, for example, if we are studying the solar system.

School geometry studies figures in 2-space and 3-space. Associating a vector to every point, we can add and subtract vectors either geometrically by putting one vector's tail at the other vector's head or algebraically by using a coordinate system. Each vector is represented by three coordinates, and vectors are added by adding the coordinates respectively. This algebraic approach easily carries over to higher dimensions. We are led into n-dimensional vector spaces and into matrix algebra, which is the machinery of calculation there.

It is one thing to talk about higher-dimensional spaces and another thing to analyze them. It certainly is puzzling to visualize even four-dimensional space, so we resort to algebra and calculus instead of visualization. Thus classical mechanics, the study of systems of particles and rigid bodies, is translated into the study of higher-dimensional geometry.

Whereas visualization in more than three dimensions seems impossible, algebra works just as well in four or five dimensions. In fact, the theory of "mechanics"—of systems of particles, constrained by rigid rods or by gravitational force or pulled back and forth by springs or rotating around an axis—demands such n-dimensional phase space.

For effective geometrical thinking we need appropriate ideas of magnitude and direction. For the magnitude of a vector in n-space, where n is any positive integer, we keep the definition that is familiar in Cartesian geometry: it is the square root of the sum of the squares of the coordinates. For direction, we define the cosine of the angle between two vectors by adapting another formula from plane and solid geometry. In elementary geometry, if two vectors (a_1, a_2, a_3) and (b_1, b_2, b_3) both have length 1, then the cosine of the angle between them is given by $a_1b_1 + a_2b_2 + a_3b_3$, an expression which is called their "inner product". So in n-space, we *define* direction by first defining the cosine of the angle between any two vectors of length 1 to be the n-dimensional inner product $a_1b_1 + \cdots + a_nb_n \cdots$ and then for a pair of arbitrary vectors in n-space "normalizing" them (dividing their inner product by their lengths) and then calculating the cosine for the normalized vectors.

Function spaces, functional analysis

This move from two and three dimensions to n dimensions, for any finite n, is the first step toward functional analysis. Now I want to continue this geometric thinking by looking at infinite-dimensional linear spaces. This is the basic setting of "functional analysis" which pervades Peter Lax's research in partial differential equations. It is directly connected to our previous conversation on the heat equation and the wave equation by means of Fourier analysis.

Linear partial differential equations are the ones we understand the best. There, smooth initial data guarantee smooth motions for all time. In that case, we have the great advantage that if any two solutions are known, then their sum and their multiples by constants are also solutions. This makes possible the method of Fourier

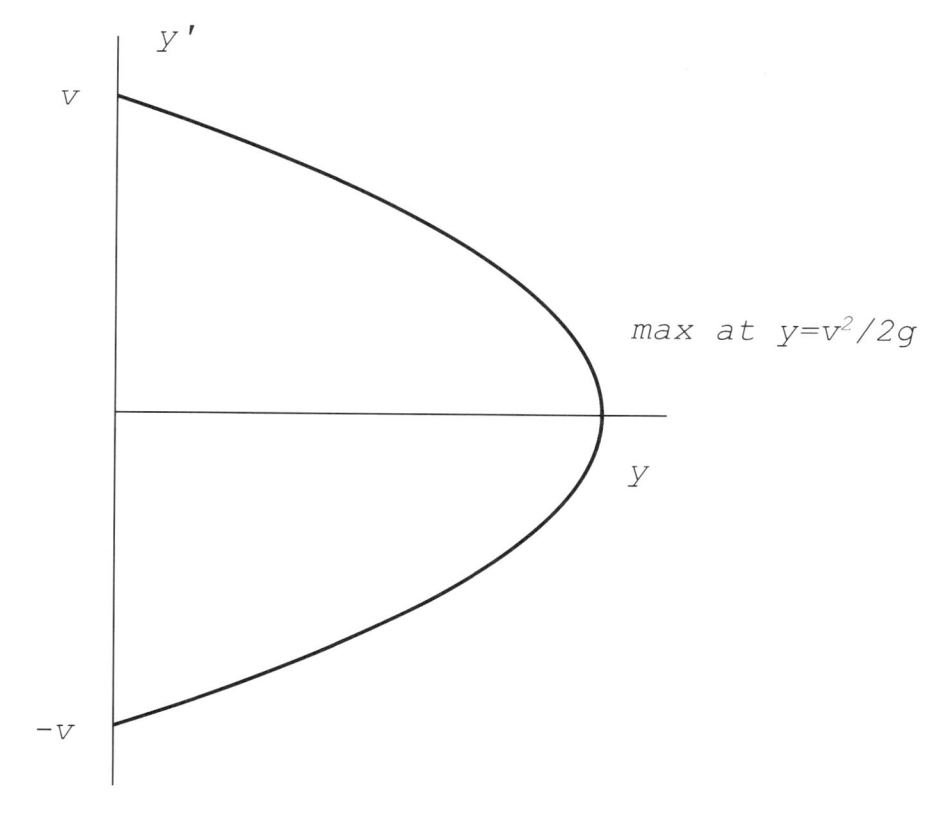

FIGURE 2. Phase plane trajectory. A stone is tossed upward from the ground with initial velocity $y' = v$. It loses speed and gains height until a maximum, and then falls, gaining downward speed, to reach velocity $y' = -v$ at ground level.

series or Fourier integrals. The solution to the general problem can be expressed as a sum of special solutions which have the form of "sine waves".

To complete the journey, we need to look at a famous formula from applied mathematics, due to Fourier. Fourier showed that any periodic function—something that repeats again and again every second, or minute, or hour and is "reasonably well-behaved", with a very broad understanding of "reasonable"—can be written as a combination of sine or cosine functions. If $f(t)$ is an odd periodic function, then there are numbers or "coefficients" c_1, c_2, c_3, etc., such that

$$f(t) = c_1 \sin t + c_2 \sin 2t + c_3 \sin 3t + \ldots.$$

There is a simple formula for the coefficients c_1, c_2, c_3, etc.

You may notice that this formula for $f(t)$ as a combination of sine functions is rather similar to the formula for any vector in some coordinate system:

$$u = c_1 e_1 + c_2 e_2 + c_3 e_3 + \cdots + c_n e_n,$$

where the vectors e_1, e_2, etc., are the "basis vectors" of the coordinate system. In calculus, where $n = 3$, they are usually called i, j, and k. The numbers c_1, c_2, etc., are the coordinates of the vector u in that coordinate system. The simple and brilliant idea of functional analysis is to think of the Fourier expansion of a function $f(t)$ as the coordinates of *a point in a very high-dimensional linear space. Infinite-dimensional, in fact.*

The first time anyone is exposed to this idea, he experiences severe "cognitive dissonance". You immediately object, "The function $f(t)$ is not just a point, it has a graph, it is a whole curve containing infinitely many points. How can you think of it as just a single point?" But there is a famous old theorem: "Each periodic function that is at all decently behaved has a convergent Fourier expansion." That means it is uniquely associated with and uniquely defined by the sequence of coefficients in its Fourier expansion. We could write this symbolically as

$$f(t) \sim c_1, c_2, c_3 \ldots .$$

From the function $f(t)$, Fourier's formulas give us the coefficients c_n. Conversely, from the coefficients we recover the function by adding the series of sine functions with these coefficients.

At first thought, it seems weird to take these coefficients as coordinates in an infinite-dimensional space. But that geometry fits amazingly well with the properties of the sines and cosines! For instance, we can define the scalar product of two functions $f(t)$ and $g(t)$, both having norm 1, to be the sum of the products of their coefficients:

$$a_1 b_1 + \cdots + a_n b_n + \cdots .$$

By analogy with Euclidean space, this sum is regarded as the cosine of an angle in "function space". It turns out to be equal to the integral of the product of the two functions $f(t)$ and $g(t)$ over the period from 0 to 2π. And by amazing good luck, the basis functions $\sin_t, \cos_t, \sin 2t, \cos 2t, \ldots$ all yield 0 when multiplied pairwise by each other and integrated. So if we take these basis functions as the coordinate vectors in this high-dimensional space, we automatically, without any special effort, have an *orthogonal* coordinate system, one where different basis vectors are mutually orthogonal—the cosine of their angle is zero; i.e., the inner product is zero. This fact alone shows conclusively that it is very natural and appropriate to identify Fourier expansions with orthogonal coordinates in an infinite-dimensional space. In fact, this is precisely the space that was named "Hilbert space" by John von Neumann.

Here I have achieved simplicity by taking the function $f(t)$ to be periodic with period 2π. The three dots at the end of the right-hand side indicate that the sum "goes on forever", that it is a convergent infinite series.

The methodologies of "functional analysis" and "Hilbert space" were among the major accomplishments of twentieth-century mathematics and included essential contributions by Hungarian, Italian, French, German, Russian, English, and American mathematicians. We do not attempt to give full credit where credit is due, but just show where Lax was coming from and how he fit in.

One essential point is the unity of geometry and algebra/analysis, going back to Descartes's discovery of the power of the algebra of his time to tackle ancient geometric puzzles and the ability of Newton to use geometric thinking to understand gravitation and light propagation.

There is no such powerful general method for nonlinear equations!

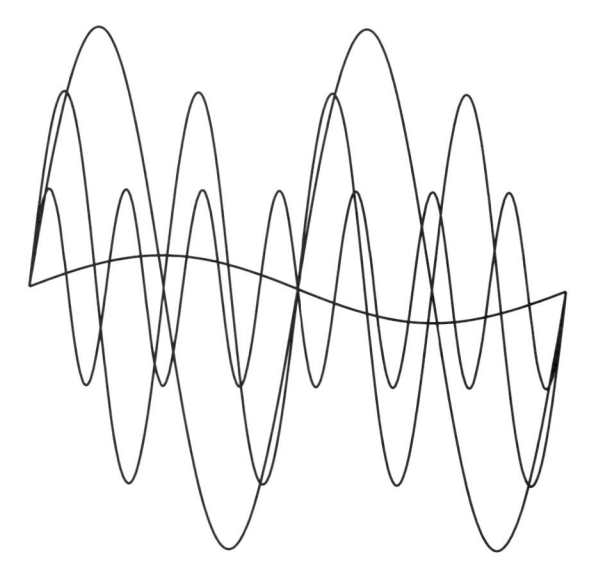

FIGURE 3. Complicated periodic functions can be represented as sums of sine waves.

Nonlinearity

Nonlinearity in mathematical analysis is a complex, fascinating issue that should interest any philosophically minded person. It is comparable to the issue of self-reference and infinite regress in computing and logic and to the issues of positive and negative feedback in engineering. It corresponds to "reflexivity" in George Soros's writings on finance and economics. The general phenomenon that arises in so many different fields can be described in the following way: something we want to know depends on something else, which depends on the very thing we want to know.

This is not just a verbal paradox, it is a real dilemma. It arises in practical situations where a decision cannot be avoided or postponed, so people have to find ways through or around it. In mathematics, that is precisely the content of nonlinear mathematics—more specifically, nonlinear differential equations.

Some of Peter Lax's most interesting work found ways to get through or around nonlinearity: first, in nonlinear conservation laws, which serve to model shock waves in compressible media, and later in soliton equations, especially the first and most famous, the Korteweg-de Vries equation of shallow water flow.

Taken literally or simplistically, "nonlinearity" is so inclusive that it is almost vacuous. Not linear? Not a straight line? Curved? Isn't almost everything really at least a little bit curved? There is an old joke—nonlinear mathematics is like "Nonelephant zoology"—it includes almost everything!

Well, it's not a trivial or pointless issue; it needs some clarifying. First of all, notice that a very short segment of an ordinary curve or trajectory is close to a bit of a straight line. This straight line is the "tangent" that just "touches" the curve at one point. The direction of the tangent line at that point defines the direction of the curve there. The tangent predicts where the curve goes from that point, but, depending on how rapidly the tangent turns as you proceed along the curve, the

FIGURE 4. The graph of this nonlinear differential equation vanishes to infinity in finite time.

tangent as a predictor may go wrong so quickly that it can lead to disaster, for instance, in weather prediction or stock market investing. Still, unreliable as it is, the linear approximation, the straight line, often is the best thing we have by which to judge where we are going as we follow along some trajectory in the economy or the weather.

So "nonlinear mathematics" could mean predicting a curving trajectory, with something more far-sighted, more sophisticated, than just the tangent line.

The term "linear" has algebraic meaning as well as geometric. It refers to functions that are additive and homogeneous. The algebraic notion of linearity is the kernel of a tremendous amount of classical mathematics.

What can happen in a nonlinear differential equation? Here's the simplest possible example.

A rocket is shooting upwards with an altitude $y(t)$ (it is y miles high after t minutes). If its velocity upward is equal to its altitude, this is a linear equation, $dy/dt = y$. The solution to this equation, the altitude, is an exponential function of the time, $y = ce^t$, where c is the altitude at time zero. The rocket zooms upward faster and faster, forever! But now suppose the velocity is equal to a slightly larger power of the altitude, say the "1 + epsilon" power, where epsilon may be quite small. Then the velocity and the altitude become infinite after a finite time! The rocket disappears!

We can be satisfied to work out the easiest example. Choose epsilon $= 1$, so that $dy/dt = y^2$, and consider the graph of the solution, with a horizontal t-axis and a vertical y-axis. Besides the "stationary points" on the horizontal t axis (y identically $= 0$), the solution to this equation (as you can easily check) is the simple function $y = 1/(c - t)$ for any constant c. The graph of this curve in the $t - y$ plane has a vertical asymptote at $t = c$. (It is one branch of a rectangular hyperbola.)

As the graph approaches that line, it shoots up to infinity; it cannot go any further to the right. The rocket disappears at time $t = c$.

The value of the constant c is determined by the altitude attained after one minute. Suppose, for example, you want it to have altitude $y = 1$ mile at time $t = 1$ minute. Then you can easily calculate that c should equal 2, and the desired solution curve is $y = 1/(2-t)$, which "blows up", becomes infinite, as t approaches 2. If $y(1) = 1$, the function $y(t)$, the solution to the differential equation, cannot be continued beyond $t = 2$. In fact, as you can easily check, at whatever time t the altitude $y(t)$ equals 1 mile, the rocket "disappears to infinity" one minute later. Any rocket rising with velocity equal to the square of its altitude (following the rule $dy/dt = y^2$) is doomed to disappear at time $t = c$ minutes, because its graph cannot cross the vertical line $t = c$. Nonlinearity can be fatal!

More interesting issues come up for functions of two independent variables—"partial differential equations". For instance, to repeat what we saw a few paragraphs above, the temperature T in a bar of metal depends on two independent variables—time t and position x. We saw that simple physical considerations about conservation of energy lead to the "heat equation"

$$dT/dt = kd^2T/dx^2.$$

On the right of the equation is the with respect to position multiplied by a coefficient k, the "diffusivity". This coefficient depends on the material the bar is made of, be it copper or aluminum. In a first course in PDE, k is supposed to be a constant. This is not too badly mistaken if the variation of the temperature T is not very great. Then the equation is linear, and we can solve it with series of sines and cosines. But in reality the diffusivity changes with the temperature, and the little letter k ought to be written as a function $k(T)$ to indicate the dependence of diffusivity on temperature. This gives us a nonlinear heat equation! The right side of the equation is now a product, a derivative of the unknown multiplied by a nonconstant function of the unknown. Superposition, adding up special solutions, doesn't work!

To do real physics of signal propagation, a similar modification must be made in the transport and wave equations and in the other equations of classical continuum mechanics. Real physics is nonlinear. We study linear oversimplifications only as a first step to get some idea of what is going on.

A standard way to cope with nonlinear equations, often taken in applied mathematics and engineering, replaces the derivatives in the equations by "finite differences". For example dy/dt is by definition the limit, as the denominator goes to zero, of the quotient of small differences in the two variables y and t. If we use the finite differences of the variables without going to the limit, the partial differential equation becomes a finite difference equation, which can be solved algebraically. If the finite differences are very small—or even *very, very* small—we can hope that the solutions we calculate algebraically will approximate the actual correct solution. But they may blow up instead! This is called "instability". Finite difference approximation must be done with care and insight.

Powerful approximation methods have been developed, mechanized, and packaged, but often they aren't accompanied by a rigorous error estimate. We can expect such an approximation to be close, but how much better it would be if by some trick, some device or other, we could actually solve a nonlinear equation! As we shall see in the next chapter, such exact solutions to certain nonlinear PDE's of continuum mechanics were discovered by Peter Lax.

CHAPTER 10

Difference schemes. Shocks. Solitons. Scattering. Lax-Milgram. Pólya's curve, etc.

In this final chapter, we take a more detailed look at several of Peter's important research contributions. We have already discussed in Chapter 4 two of his early famous papers: the Lax-Richtmyer paper with the equivalence principle of numerical analysis and the paper on diffraction that would be generalized to "Fourier integral operators". Now we give a general overview of four major continuing themes: shock waves, finite difference approximation, solitons, and scattering. In addition, we look at the Lax-Milgram lemma, the differentiability of Pólya's curve, and a long-standing conjecture on hyperbolic algebra.

Difference equations and computing

Peter's contribution to computational mathematics is both theoretical and practical. He first became fascinated with the potentialities of high-speed computers in 1945 as a corporal in the Manhattan Project, working with John von Neumann. After finishing his Ph.D. at NYU, he met von Neumann again in 1949 when he returned to Los Alamos for a year. "When we met again he was always available. When I latched on to some of his ideas like von Neumann criterion, stability, shock capturing, he was interested in my ideas and I profited from his ideas."

(We say more about this in the next section, on shock waves.)

Peter continued visiting Los Alamos every summer for several years.

We have discussed his best-known contribution to the theory of difference schemes, the Lax-Richtmyer equivalence theorem, in Chapter 4. On the practical level, two very popular and well-known algorithms for computing shocks bear Peter's name. These schemes serve both as benchmark tests for other numerical techniques and as starting points for theoretical analysis. One of them is "Lax-Friedrichs". The Russian mathematician Olga Oleinik used the Lax-Friedrichs scheme in her constructive proof of the existence and uniqueness of solutions of the inviscid Burgers' equation. Kurt Friedrichs was the director of Peter's Ph.D. thesis. He was not interested in doing computing, although a numerical algorithm bears his name. Friedrichs was interested in positive schemes merely as tools for proving existence.

The other algorithm is "Lax-Wendroff". Burt Wendroff is a New Mexican like me, an old friend. Lax-Wendroff is more accurate than Lax-Friedrichs and is very successful at "shock capturing", the approach to computing shock waves first proposed by John von Neumann. Another useful result is the Lax-Wendroff theorem: "If a numerical scheme for a nonlinear hyperbolic conservation law converges strongly to a limit, then the limit is a solution of the equation."

Numerical methods have difficulty resolving shocks, and it's hard to prove that a calculated candidate for an approximate solution is actually close to the physical solution. The Lax-Wendroff work, like several other of Peter's contributions to applied mathematics, succeeded by dint of insights from abstract mathematics applied to specific issues of calculation.

"Wendroff and I investigated a particular difference equation in conservation form. We pointed out that in order for the limit to satisfy the differential conservation laws, then in contrast to the linear case the limit has to be in the strong, not the weak sense."

Peter reviewed von Neumann's early calculations and conjectures. "Von Neumann saw the importance of stability, and furthermore he was able to analyze it. His analysis, called the von Neumann criterion, is rigorous for equations with constant coefficients, where Fourier analysis tells the story. He conjectured that even for nonlinear equations, if you apply his criterion locally by freezing the coefficients, and the scheme passes his test at each point, then the whole scheme is stable. He was right, by and large, but you must put down a few minor but important restrictions.

Peter Lax and Burt Wendroff

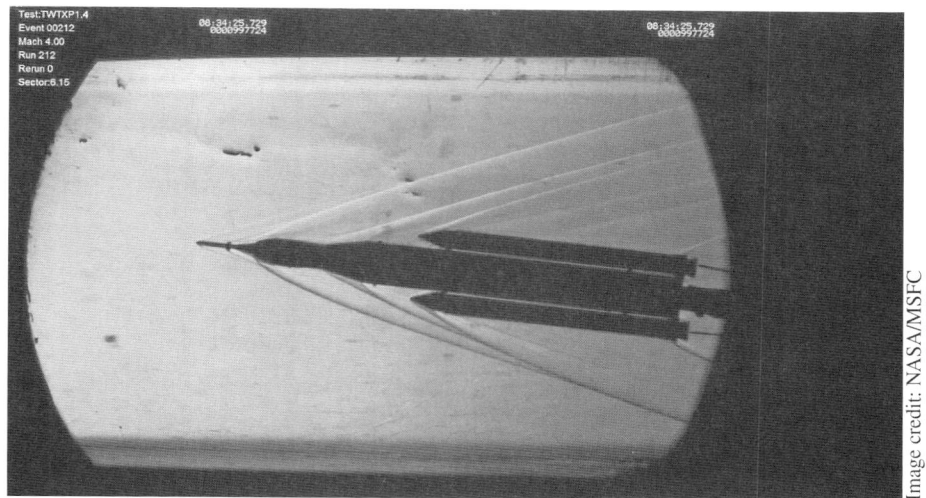

Shock waves

Image credit: NASA/MSFC

"That was the place where shock capturing was first introduced. Von Neumann used a dispersive method, and the solution that he computed was oscillatory. This did not bother him. He and Richtmyer introduced artificial viscosity to suppress the oscillations."

Shock waves

What can it mean to talk about a "discontinuous solution" to a differential equation, an equation about the derivatives of the unknown function? Well, there certainly are shocks—sonic booms, for example—in the atmosphere. So, mathematically, these must be discontinuous solutions of the equations of fluid dynamics. Making sense of this is a challenging mathematical problem. The hard part isn't saying what you mean by the existence of such solutions, but finding the right supplementary condition to pick out the unique solution which is physically correct.

For an incisive introduction to the mathematics of shock waves, one cannot do better than Peter's own article, "The formation and decay of shock waves", which won two awards for expository writing from the Mathematical Association of America: the Lester R. Ford Award in 1973 and the Chauvenet Prize in 1974. It appeared in *The American Mathematical Monthly* in 1972 (vol. 79, pp. 227-241). It describes the origin of the equations governing the propagation of shock waves, some of the striking phenomena, and a few of the mathematical tools used to analyze them.

The most familiar example of a shock wave is a clap of thunder. When an airplane overhead "breaks the sound barrier", you feel and hear a "sonic boom", a shock wave. When the speed of the plane becomes greater than the speed of sound, it overtakes and passes the sound waves it has previously generated. The sound waves pile on top of each other, creating a "shock"—a sudden discontinuity in air pressure.

Planes started breaking the sound barrier during World War II, and shock waves became a military concern. A major study was made by Richard Courant and Kurt Friedrichs on behalf of the U.S. military. (A few years later, both Courant and Friedrichs would become Peter Lax's mentors.) Their book *Supersonic Flow and*

Kurt Otto Friedrichs at work with Richard Courant, 1965

Shock Waves was edited by Cathleen Morawetz when she was a graduate student at NYU. The book shows pictures of an airtight cylinder, a "shock tube". Inside the cylinder a piston can drive forward, like the piston inside a cylinder in your automobile engine. In front of the moving piston moves a shock—a compression wave, a moving discontinuity of air pressure.

To help visualize the physics, Friedrichs and Courant write, "Assume a long column of equally spaced cars traveling at supersonic speed, which strikes an unanticipated obstacle that suddenly brings the first car to a full stop. The second will press close to the first and stop, then the third will be abruptly stopped by the second, and so on. The point separating the stopped cars from the moving cars obviously represents a receding shock front. A forward-facing shock front impinging on a zone of rest is represented by the phenomenon of a column of fast moving cars pounding against a row of widely spaced parked cars and setting them in motion." (*Supersonic Flow and Shock Waves*, pages 129-130). In the shock tube, something analogous is happening to the air molecules ahead of the piston.

At the Manhattan Project shock waves were a major concern. First of all, the nuclear explosion is created by an implosion wave driving separate pieces of plutonium into each other, which is a shock wave. After the bomb explodes, the first big effect is a tremendous heat wave, with X-rays and other deadly radiation. Then the bomb blast sends out a terrific pressure wave, a shock, that knocks down buildings that have just been set on fire. So Los Alamos became a main center for studying shock waves.

Mathematically, they are interesting because the normal numerical approximation procedure for calculating the solution of a partial differential equation breaks down at discontinuities—shocks. Finding out how shocks are created and how they propagate is a fascinating mathematical problem. It is said that of all the mathematical problems raised by the bomb project, shock waves were Johnny von Neumann's favorite. And in this work Peter Lax collaborated and was inspired.

A shock can be understood as the result of a faster signal overtaking a slower one. If the information carried by the two signals is different, they are incompatible when they intersect; they contradict each other, giving rise to a discontiniuity.

Shock waves in the air are part of the theory of "compressible flow". (In contrast, "incompressible flow" governs water waves.) The nonlinear conservation laws governing compressible flow were derived in the eighteenth century by Leonhard Euler. To describe the motion of air, we need five "state variables": pressure, density, and three components of velocity. The state variables are governed by an equation of state and three basic physical principles: conservation of mass, conservation of momentum, and conservation of energy. Assuming that mass isn't being created or destroyed, any loss or gain of mass inside any region must equal the net mass passing through the boundary of that region. This equality is stated mathematically by an equation between a volume integral over the interior of the region and a surface integral around the boundary of that region. Two theorems of calculus, the divergence theorem and Stokes' theorem, convert the surface integrals to volume integrals. If the state variables are smooth continuous functions, these equalities between integrals imply equalities between derivatives—"partial differential equations". The equations are nonlinear, because the relationships between the derivatives of the state variables depend on the state variables themselves.

Even if the density, pressure, and velocity start out perfectly regular, the flows can develop shocks, or discontinuities.

Peter Lax has written, "When in the middle of the nineteenth century it was observed that solutions of equations of compressible fluids developed discontinuities, it rang an alarm bell. After all, the equations seemed so plausible! It was a genuine crisis. People like George Stokes were baffled. The right answer was found by no less a mathematician than Bernhard Riemann."

In 1860 Riemann studied a shock tube where a thin membrane divides a cylinder into two sections at two different constant pressures. The membrane is pierced, and two waves move out from the membrane: a compression wave into the lower pressure chamber, and a rarefaction wave into the higher pressure chamber. Both waves are fronts of discontinuous air pressure. The differential equations of compressible flow are expressed in terms of the derivatives of the density and pressure and are not meaningful at the shock front, where pressure and density are discontinuous. What are the speeds of the two shock waves, and what are the air pressures in different parts of the shock tube? Peter Lax wrote, "Riemann observed that the physical conservation laws are integral equations. The integral equations can have perfectly good discontinuous solutions, which are not admissible as solutions of the differential equation. Riemann went further and derived the laws of propagation of these discontinuities. Today they are called the Rankine-Hugoniot relations, but really they should be called Riemann-Rankine-Hugoniot relations. Riemann made a mistake, but the mistake was in thermodynamics, which was not yet a science at this time."

It is instructive to work out in detail this simple example of a nonlinear conservation law in one dimension using only elementary calculus and a space-time

diagram:
$$u_t + d(u^2/2)/dx = 0,$$
which can be written as

(A) $\quad u_t + uu_x = 0 \quad$ which is equivalent to (A$'$)$u = -u_t/u_x$.

Call the dependent variable $u(x,t)$ "pressure". The term du/dt is the rate of change of pressure with respect to time, and the term du/dx is its spatial rate of change. The equation is nonlinear because it contains the product of the pressure u times its spatial derivative du/dx.

Let's use the horizontal axis as the x-axis and the vertical axis as the time t-axis. (This is the reverse of the usual choice in elementary calculus.) Then the slope of a curve in these coordinates will be the derivative dt/dx. Corresponding to any value c of the pressure u at some times t and positions x, the points in the space-time diagram defined by the equation $u(x,t) = c$ trace out a curve. This curve through the space-times where the pressure is equal to c is called an "isobar". The isobar through any initial point on the x-axis tells us how to move right or left as time goes on so as to keep our pressure constant.

To solve equation (A), we will find the isobars. In fact, we will see that the isobar $u(x,t) = c$ is a straight line with slope $1/c$.

In order to show this, we use equation (A$'$) and the chain rule of calculus to calculate dx/dt, which is the reciprocal of the slope dt/dx.

On the isobar, u is a function of one variable, t, and it is constant, so its derivative $du/dt = 0$. On the other hand, the value of u on the isobar is the result of restricting a function of two variables, $u(x,t)$, to the curve $x = x(t)$:
$$u(t) = u(x(t), t).$$

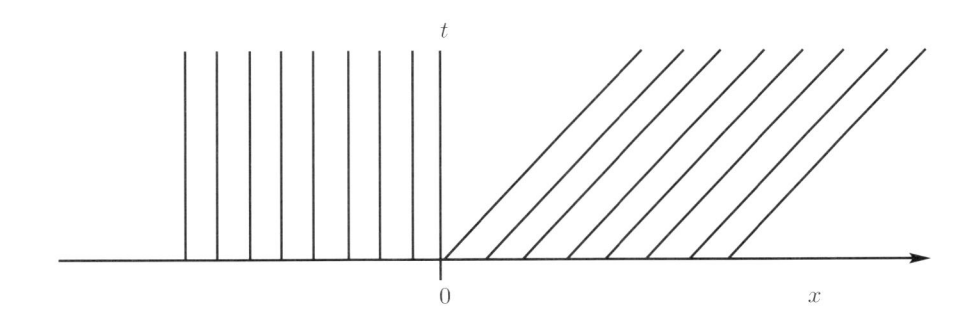

The solution is under-determined by the initial data.

Using the chain rule of elementary calculus, differentiate $u(t) = u(x(t), t)$ with respect to t:
$$\text{on the isobar, } du/dt = (u_x)(dx/dt) + u_t = 0.$$
Solve this equation for dx/dt:

(B) $\qquad\qquad\qquad dx/dt = -u_t/u_x.$

Now compare equations A$'$ and B: together they say that $dx/dt = u$. On the isobar, $u = c$, so $dx/dt = c$, and its reciprocal, dt/dx, the slope, equals the constant $1/c$. A curve with constant slope is a straight line, so we have shown that the isobars are straight lines with slope $1/c$, as we claimed.

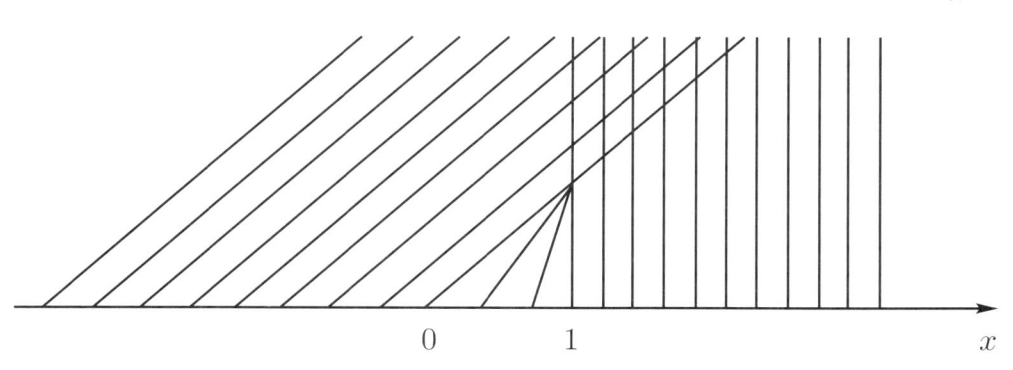

The solution is over-determined by the initial data. Contradiction!

But how do we assign the right constant pressure c to any isobar? We can do so for any isobar that intersects the initial line $t = 0$, because there we have the known initial values of the pressure. At any point x_0 at $t = 0$ the value $u(x_0)$ is assigned initially, and on the ray coming out of that point, with slope $1/u(x_0)$, u has the constant value $u(x,t) = u(x_0)$. These rays will cover some part of the upper half of the space-time diagram, where $t > 0$. They may leave part of it uncovered, as we see in Figure 1, or they may intersect, as we see in Figure 2, and leave us in doubt which ray to continue beyond the point of intersection, since they cannot both be valid there.

As we look to the right or left along the initial line, the slopes of the rays vary according to the assigned initial values of u. We see immediately that if, as x_0 increases (as we move to the right), *the slopes of the rays are decreasing continuously (i.e. the wave speeds, their reciprocals, are increasing)*, then the rays fill out the upper half plane. But if as we move to the right *the slopes are increasing and the speeds are decreasing*, the rays coming out of the initial line will intersect. The intersecting rays will carry incompatible constant values of u, and so at such points of intersection our PDE with its initial values is inconsistent—contradictory. *Shocks*! There is just no solution! However, if this simple equation is meant to model a possible physical system, then, "after all, the world does not come to an end," as Peter has written. Our mathematical equation has broken down—that means it no longer works as a model of the physics. We have to redesign the model, find the mathematical description which accounts for the shock, and what happens after the shock has passed by.

"When I started to work on conservation laws I was very much influenced by two papers: Eberhard Hopf's on the viscous limit of Burgers' equation, and von Neumann and Richtmyer's on artificial viscosity. After looking at these examples, I could see what the general theory should look like. In my very first paper I was able to prove the rather remarkable compactness of solutions. I did it by using weak convergence versus strong convergence."

In a 1957 paper, "Hyperbolic systems of conservation laws II", he gave a general solution to the Riemann type of initial value problem, not just for the classical equations of compressible flow but for a general system of n conservation laws. In addition, he gave the correct formulation of the classical entropy condition to determine uniquely the right continuation past a shock.

For a single conservation law, in one space dimension, Peter Lax found an explicit solution formula with a general nonlinearity and a general initial state. His formula generalizes a formula of Hopf for a quadratic conservation law. Lax writes the solution as $b((x - y_0)/t)$, where the function $b(.)$ is defined as the value of the initial function $u(x)$ at the point x where a certain explicit elementary function $g(s)$, parametrized by u, is maximized, and y_0 is the point where another explicit elementary function is minimized. It is quite extraordinary for a whole class of nonlinear partial differential equations of major physical interest to be solved by an explicit formula.

For a system of several conservation laws in one space dimension, he found a simple explicit solution, provided that the initial state is two different constant states, one for positive and one for negative x, with a jump discontinuity at $x = 0$. This includes the physical problem studied by Riemann in 1860. Peter used $n - 2$ intermediate constant states, connected in succession by either shock waves, contact discontinuities, or "simple waves" (waves following straight-line trajectories in space-time). The "jumps", or discontinuities across the shock waves, are determined by Peter's entropy condition, consistent with the classical conditions of Rankine and Hugoniot.

The bulk of that paper developed the "entropy condition", which solves the problem of choosing the unique correct continuation across a shock. The condition is an inequality between two derivatives which states that across a shock front entropy must be constant or increasing. In the physical case of compressible flow, $n = 3$: pressure, density, entropy.

"I knew that shocks and contacts were very different in fluid dynamics, and I wanted to put this into a general context. A surprise was that in general, the deviation of weak shocks from the rarefaction rate is of third order. In fluid dynamics, that statement says that entropy changes in third-order shock steps. And that theorem turned out to be true, not just in fluid dynamics, but as a general algebraic fact about systems of conservation laws! That surprised me."

Peter's solution of this general Riemann problem is now a cornerstone in the theory of hyperbolic conservation laws. It stimulated extensive further research into different entropy conditions applicable to other systems. The fundamental existence result proved by Glimm for the general initial-value problem uses the Lax theorem as a building block.

A system of nonlinear conservation laws, propagating in three-dimensional space, constitutes to this day a very difficult mathematical problem.

In the presence of a shock, the solution to a wave problem can be either underdetermined (there are points in space-time where the initial data do not determine the solution) or overdetermined (where the initial data may be inconsistent). To overcome this blockage, there are two different approaches. One, called "artificial viscosity", adds a second-order term to the equation, introducing a little "viscosity" and causing some loss of energy. The shock is smoothed out, but is the solution of this modified equation really close to the original discontinuous wave? The other method is to replace the differential equations by finite difference equations.

"Shock tracking is an important problem, but keeping track of the location of the shocks is laborious and hard to program. It's difficult to know that a new shock is about to start. And in more than one space dimension, tracking shocks is impossible. Von Neumann came up with the idea of shock *capturing*: don't keep track of a shock as a distinct discontinuity, but instead regard it as a rapid transition. The great contribution of von Neumann was disregarding the discontinuity. But he gave no theoretical justification.

"I pointed out that if you write the equations in conservation form, then you can define what it means for a discontinuous function to be a solution. It's the analogue for nonlinear conservation laws, of the weak solution for linear hyperbolic equations. I further pointed out that if the difference equations are also in conservation form, then their limit will be a weak solution of the differential conservation law.

"My contribution was seeing that you have to set up the difference equation in a conservation form consistent with the original conservation laws. That was a very useful observation. I had tried it out and it worked. The calculations are in 1954 in the *Communications on Pure and Applied Mathematics*. In shock capturing, one can actually prove that when a difference equation in conservation form converges strongly, then, in the weak sense, its limit satisfies the differential equation. That's not limited to the conservation laws of fluid dynamics; it's true for all hyperbolic conservation laws."

The selection principle known as the entropy condition was one of Lax's major contributions. To this day there is no complete theory for shock flow in two or three dimensions.

"You want greater accuracy, but even more you want greater resolution. I defined a concept of resolution. If you apply a difference method to a set of initial value problems and look at the states into which it develops after a fixed time, the surprise is that this set is much smaller for nonlinear equations. Wave information is actually destroyed, and it's a much smaller set. The measure of the relevant set is called its entropy or capacity. A method has a proper resolving power if the measure of this set is comparable to the measure of the set of exact solutions. If it's very much smaller, it cannot resolve. First-order methods' resolution is too low, many details are just washed out. Second-order methods have better resolution. Could it be that methods higher than second order have too much resolution, more resolution than you need?"

He says, "It's quite amusing that while the number of people willing to work on the equations of compressible fluids was always limited, the number of people willing to work on a more general problem, hyperbolic systems of conservation laws, was much larger!"

Recently it has been proposed that shock waves may play a role in cosmology and the large-scale structure of the universe. In the standard model of cosmology the equations of compressible flow are coupled to the equations of general relativity in a nonlinear system of partial differential equations that can support shock waves. Joel Smoller of the University of Michigan and Blake Temple of the University of California, Davis, have incorporated a shock wave into the Standard Model of Cosmology, supplying the simplest physically plausible finite-mass cutoff for the Big Bang theory. In their model, the universe today lies within a time-reversed black hole, from which it emerges, after the shock wave comes into view, and finally, as a ball, expands into empty space. They explore the possibility that "dark energy" (the physical interpretation of the anomalous acceleration of the galaxies) may come from a self-similar wave formed during the radiation phase of the Big Bang. (When the nonlinear mechanisms of pure radiation are included, shock wave theory

Soliton experiment on Union Canal, near Heriot-Watt University, Edinburgh, Scotland, August 1995

predicts decay to self-similar waves.) Thus shock wave theory gives a cosmos far different from standard cosmology.

Solitons

Solitons are nonlinear waves with a mysterious particle-like property: they pass through each other while preserving their shape. The soliton was first observed as a water wave running down a canal. Then interacting multi-solitons were discovered in computer experiments by Martin Kruskal and his collaborators. They satisfy a certain nonlinear PDE, the Korteweg-de Vries equation (KdV). Lax invented a simple method of analyzing KdV, "the Lax Pair". Given an operator K depending nonlinearly on a parameter u, a Lax pair for K is a pair of linear operators $A(u)$ and $B(u)$ whose commutator $AB - BA = K$. This device not only worked for the KdV equation, it revealed solitons in a whole host of other nonlinear equations of physics and geometry.

Solitons were discovered first in 1834, and then again in 1967. On an August day in 1834, next to the Union Canal at Hermiston near Edinburgh, Scotland, a young engineer named John Scott Russell was watching a boat being pulled rapidly along the channel. The rope pulling the boat broke, and the boat suddenly stopped. "The water in the channel accumulated round the prow of the vessel in a state of violent agitation. Then suddenly left it behind, and rolled forward with great velocity, assuming the form of a large solitary, smooth and well-defined heap of water, which continued its course along the channel without change of form or diminution of speed. I attempted to follow the wave on foot, but finding its motion too rapid, I got instantly on horseback and overtook it in a few minutes still rolling on at a rate of some eight or nine miles an hour, preserving its original figure some thirty feet long and a foot to a foot and a half in height. Its height gradually diminished, and after a chase of one or two miles I lost it in the windings of the channel. Such, in the month of August 1834, was my first chance interview with that singular and beautiful phenomenon which I have called the Wave of Translation."

Scott Russell studied "Waves of Translation" or solitary waves for several years with controlled laboratory experiments. He showed that large-amplitude solitary waves in a channel move faster than small ones—a nonlinear effect. All his life, Russell remained convinced that his solitary wave was of fundamental importance, as a self-sufficient dynamic entity, a "thing" displaying many properties of a particle. Other scientists, including Airy and Stokes, were skeptical of his observations.

In 1872 a French mathematician, Joseph Boussinesq, studying Russell's "wave of translation" found a new approximation for water waves. Because waves spread horizontally rather than vertically, Boussinesq eliminated the vertical coordinate from the equations while still keeping some influence of the vertical structure of the flow.

Boussinesq's equations let the wave travel simultaneously in two opposite directions. Then in 1895 the Dutch physicist Diederik Johannes Kortewegwith his student Gustav de Vries obtained a simpler equation for waves traveling in only one direction. (Korteweg (1848–1941) was a well-known Dutch scientist. His obituaries do not mention this paper, which was "in essence a fragment of the dissertation of Gustav de Vries, completed under the guidance of Korteweg and defended on December 1, 1894. De Vries was a high school teacher and almost nothing is known about him" [*Mathematics as Metaphor*, Yuri I. Manin, American Mathematical Society, 2007, p. 102]).

Their equation is
$$u_t = -uu_x - u_{xxx},$$
now called the Korteweg-de Vries or KdV equation. The variable u is the distance from the surface of water to the bottom.

Korteweg and de Vries found a formula for a solitary traveling wave like the strange wave Russell had followed on horseback. Using the "hyperbolic secant" function (abbreviated sech x) which equals $2/(e^x + e^{-x})$, their solitary wave can be written as
$$u(x,t) = 3c \operatorname{sech}^2[\sqrt{c}(x-ct)/2].$$
They also found periodic solutions to their equation: "cnoidal waves", which are approximate solutions of the Boussinesq equation. But they were unable to produce general solutions. Their work and Russell's observations then fell into obscurity until the early 1960s, when it became one of the most active, useful areas in recent mathematics.

Lax writes in his article "The flowering of applied math...": "Even more unexpected was Kruskal and Zabusky's discovery of solitons, their curious interaction with each other, and their relation to the existence of infinitely many conserved quantities, and the complete integrability of systems with soliton-like structures. It is astonishing that there are so many completely integrable systems—KdV, sine-Gordon, nonlinear Schrödinger, Toda, etc.—unrecognized as such in the classical days of Hamiltonian mechanics. It is doubly astonishing that they all have a measure of physical significance. That one of them, the Kadomtsev-Petviashvili equation, arising in the study of water waves, has led Dubrovin, Arbarello, DeConcini and Shiota to a solution of Schottky's classical problem of characterizing Riemann matrices in the theory of Riemann surfaces is truly mind-boggling."

144 CHAPTER 10. DIFFERENCE SCHEMES. SHOCKS. SOLITONS.

Martin Kruskal, 1993 (1925-2006)

To tell this story, we must detour to Los Alamos in 1951. When von Neumann's MANIAC was completed in 1951, Enrico Fermi and Stanisław Ulam were eager to use it to explore nonlinear physics and mathematics. Fermi proposed a numerical study of something that he expected to be very simple: a chain of weights connected by nonlinear springs. Motion would be allowed only along the line of the chain, with no loss of energy to friction or internal heating. Without losing energy, the linked weights would oscillate forever. The restoring force in the springs would be nonlinear, the third or fourth power of the displacement.

They started their numerical model running in a simple oscillation mode. If the spring forces were linear functions, that initial mode would continue indefinitely. But with nonlinear springs, different modes of oscillation become excited. Fermi, Pasta, and Ulam expected the system to "thermalize", meaning that eventually the energy of the vibrating masses would be partitioned equally among all possible modes of oscillation. Indeed it seemed, after the computer had run for a while, that its output had reached the expected state of "noise". The researchers went out for lunch, but they had forgotten to turn the machine off! So one of them returned to the lab, where he saw something incredible—gone was the state that they had thought was a final equilibrium. The system was back in its initial state! It was as though a bottle of ink dropped into a swimming pool had spread all through the pool and then reassembled into the original single blob of ink! Or as if a set of billiard balls, after being scattered around the table from their initial triangle setup, later reassembled into the original triangle!

What does this Fermi-Pasta-Ulam (FPU) experiment have to do with solitary water waves and the Korteweg-de Vries equation? The connection was made by Norman Zabusky and Martin Kruskal. They modified the FPU model by shrinking the masses and springs down to infinitesimal size to produce a continuous line of deformable material. This continuum would obey a certain partial differential equation, and this equation, after still another transformation, becomes KdV! Thus the periodicity of the nonlinear spring-mass system is connected to the periodicity of solutions of the KdV water-wave equation!

To analyze the KdV equation, Kruskal and Zabusky had a great advantage over Korteveg and de Vries—high-speed computing! They could actually watch the evolution of the solutions to KdV as a movie, and observed with amazement the existence of multi-soliton solutions, where separate solitary waves travel toward each other, interact, and then pass on, with their original shapes unaffected by the interaction! They exchange relative positions and then refocus almost exactly at another location in space. In this sense, they're like particles. (Hence the name "soliton" to sound like boson or fermion or hadron.) Such a phenomenon in a nonlinear equation was previously unimagined.

These insights from computation made possible the ultimate success of mathematical analysis, including the inverse scattering solutions and the Lax Pair reformulation and methodology. As von Neumann had foreseen, the equation could be explored numerically even before it could be deciphered analytically. The computer was crucial twice: first, in the Fermi-Pasta-Ulam recurrence phenomenon and then in the KdV soliton and multi-soliton phenomenon.

In 1967 Gardner, Greene, Kruskal, and Miura discovered an ingenious method, which they called the inverse scattering transform, for solving KdV. They linearize the equation, in a new sense of the term, by relating it to the linear Schrödinger equation. Important quantities called "eigenvalues" of this Schrödinger equation remain unchanged under the KdV evolution. They are constant in time and are called "conserved quantities" and "scattering data". Kruskal et al. introduce a Gelfand-Levitan integral equation associated to these scattering data. From the solution of this integral equation, they obtain the solution to KdV.

A "direct" scattering computation derives scattering data from solutions of the partial differential equation. Kruskal et al. find the scattering data first and use it to solve the KdV equation, so theirs is an "inverse scattering" method.

Peter Lax discovered the inner structural explanation for the striking behavior of the KdV equation. "Since I was a good friend of Kruskal, I learned early about his discoveries, and that started me thinking. There are infinitely many conserved quantities, and I asked myself: How can you generate all at once an infinity of conserved quantities? I thought that they would be the eigenvalues of an operator depending on the solution of KdV that remains similar as t changes."

"Remaining similar as t changes" means intuitively "keeping the same shape while spinning around," like a football spinning around while the lengths of its three principal axes remain constant. (The three axis lengths of the football are the "eigenvalues" of an associated matrix.) In high-dimensional function space, such shape-preserving motions are called "orthogonal transformations". Two transformations L and M are called "similar" if an orthogonal transformation U connects them by

$$L = U^{-1}MU.$$

From this formula, Peter derived, by a few lines of simple calculation, that an equation of the form

$$u_t = K(u),$$

where K is a *nonlinear* differential or difference operator such as appears on the right side of KdV, will be "completely integrable" if linear operators A and B, depending parametrically on the parameter function u, can be found such that

$$K = BA - BA.$$

This simple equation is now called "the Lax equation", and the two operators A and B are called a "Lax pair". B is required to be antisymmetric.

In fact, for the Korteweg-de Vries equation in the form

$$u_t = uu_x + u_{xxx}$$

he found such a pair! Taking A as the Schrödinger operator, depending on a parameter $u(x)$ as follows:

$$A = -(d/dx)^2 + u(x),$$

Lax found the partner for this operator, also depending on the parameter $u(x)$:

$$B = -4(d/dx)^3 + 3(u(x)d/dx + d/dx(u(x))).$$

Indeed, an easy elementary calculation shows that $AB - BA$ is equal to

$$u_{xxx} - 6uu_x,$$

which is reducible to the right side of KdV!

From the Lax equation, not only does it follow that the eigenvalues of this A will be integrals of KdV, it follows *furthermore* that for *any other* equation with a nonlinearity that can be written in the form of the Lax equation, the eigenvalues of A will be integrals of *that* nonlinear equation and, therefore, that other nonlinear equation is completely integrable by the inverse scattering method. Solitons must also appear in the other nonlinear equations where this inner structural feature is present!

It became imperative to look for Lax pairs for other differential equations of mathematical physics. There was a big spurt of activity. Startling new connections between several different areas of mathematics were uncovered. A whole collection of nonlinear equations of geometry and physics turned out to be amenable to Lax's analysis: the Sine-Gordon equation, the Kadomtsev-Lukashvili equation, the nonlinear Schrödinger equation, the Toda lattice are the best-known examples. The insight from functional analysis which Peter Lax brought to this scene made the amazing properties of the KdV equation natural or understandable and put the soliton in a general context, not limited just to KdV.

The Sine-Gordon equation is a particularly appealing example of a nonlinear equation where Lax's method is successful. This equation was studied in the nineteenth century in connection with surfaces of constant negative curvature. Its soliton solutions attracted a lot of attention in the 1970s. They can be visualized as "kinks" and "antikinks" sliding around an elastic band.

In space-time coordinates (x, t), the Sine-Gordon equation reads:

$$\varphi_{tt} - \varphi_{xx} + \sin\varphi = 0.$$

In "light-cone coordinates" (u, v), where

$$u = \frac{x+t}{2}, \quad v = \frac{x-t}{2},$$

the equation becomes

$$\varphi_{uv} = \sin\varphi.$$

This is the form in which the equation appeared in the nineteenth century in connection with the surface of constant negative curvature called the "pseudosphere". In an appropriate coordinate system, a compatibility condition between the first and second fundamental forms results in this equation. (The name "Sine-Gordon" was adopted as a pun on the "Klein-Gordon" equation

$$u_{tt} - u_{xx} + u = 0.)$$

The Sine-Gordon equation, like Korteweg-de Vries, has soliton and multi-soliton solutions. The 1-soliton solutions are given by:

$$\varphi_{\text{soliton}}(x, t) := 4\arctan e^{m\gamma(x-vt)+\delta}$$

where

$$\gamma^2 = \frac{1}{1-v^2}.$$

The 1-soliton solution associated with the positive square root for γ is called a kink and represents a twist in the variable φ which takes the system from one solution $\varphi = 0$ to an adjacent one with $\varphi = 2\pi$. The states $\phi = 0 \pmod{2\pi}$ are constant solutions of zero energy, known as vacuum states. The 1-soliton solution where we take the negative square root for γ is called an antikink.

The 1-soliton solutions can be visualized with the use of the elastic-ribbon Sine-Gordon model discussed by Dodd and co-workers. Take a clockwise twist of the elastic ribbon to be a kink. The counterclockwise twist will be an antikink. The 2-soliton solutions of the Sine-Gordon equation show features similar to the KdV solitons. The traveling Sine-Gordon kinks and/or antikinks pass through each other as if perfectly permeable; the only observed effect is a phase shift. Since the colliding solitons recover their velocity and shape, such an interaction is called an elastic collision. Interesting 2-soliton solutions arise from a coupled kink-antikink behavior known as a "breather". There are three types of breathers: standing breather, traveling large-amplitude breather, and traveling small-amplitude breather. The Sine-Gordon equation also plays a role in quantum field theory.

Other completely integrable nonlinear equations arise, like KdV, from simplifying the Boussinesq equations for water waves. We can mention the Kadomtsev-Petviashvili equation for wave propagation in two horizontal dimensions and the nonlinear Schrödinger equation (NLS equation) for the complex valued amplitude of narrow-band waves (slowly modulated waves). Lax pairs enabled mathematicians to solve both of these.

The nonlinear Schrödinger equation can be solved with the inverse scattering transform. The corresponding linear system of equations is known as the Zakharov-Shabat system. The Schrödinger equation arises as a compatibility condition of the Zakharov-Shabat system. Starting from the trivial solution and iterating, one obtains the solutions with n solitons.

In optics, the nonlinear Schrödinger equation occurs in the Manakov system, a model of wave propagation in fiber optics. The function ψ represents a wave, and the nonlinear Schrödinger equation describes the propagation of the wave through a nonlinear medium. The equation models many nonlinearity effects in a fiber, including but not limited to self-phase modulation, four-wave mixing, second harmonic generation, stimulated Raman scattering, etc.

The nonlinear Schrödinger equation occurs in water waves

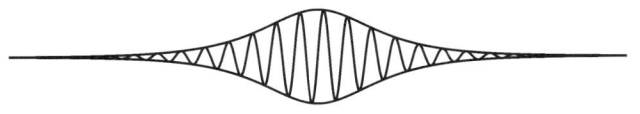

Sech envelopes

The higher-frequency inner line on this graph represents surface waves on deep water. The graph shows their envelope, which is a soliton and a hyperbolic secant function. The evolution of the envelope of modulated wave groups is governed by the nonlinear Schrödinger equation.

The Toda lattice, named after Morikazu Toda, is a simple model for a one-dimensional crystal in solid-state physics. It is given by a chain of particles with nearest-neighbor interaction described by the equations of motion

$$\frac{d}{dt}p(n,t) = e^{-(q(n,t)-q(n-1,t))} - e^{-(q(n+1,t)-q(n,t))},$$
$$\frac{d}{dt}q(n,t) = p(n,t),$$

where $q(n,t)$ is the displacement of the n-th particle from its equilibrium position, and $p(n,t)$ is its momentum (mass $m=1$). The Toda lattice is a prototypical example of a completely integrable system with soliton solutions. To see this, one uses Flaschka's variables

$$a(n,t) = \frac{1}{2}e^{-(q(n+1,t)-q(n,t))/2}, \qquad b(n,t) = -\frac{1}{2}p(n,t)$$

Ralph Phillips (1913-1998)

such that the Toda lattice reads
$$\dot{a}(n,t) = a(n,t)(b(n+1,t) - b(n,t)),$$
$$\dot{b}(n,t) = 2(a(n,t)^2 - a(n-1,t)^2).$$
Then one can verify that the Toda lattice is equivalent to the Lax equation where $[L,P] = LP - PL$ is the commutator of two operators. The operators L and P, the Lax pair, are linear operators in the Hilbert space of square summable sequences given by
$$L(t)f(n) = a(n,t)f(n+1) + a(n-1,t)f(n-1) + b(n,t)f(n),$$
$$P(t)f(n) = a(n,t)f(n+1) - a(n-1,t)f(n-1).$$
In particular, the Toda lattice can be solved by virtue of the inverse scattering transform for the Jacobi operator L. The main result implies that sufficiently fast-decaying initial conditions will split, asymptotically for large t, into a sum of solitons and a decaying dispersive part.

Peter was asked, "How important do you think this theory is for application?"

"It is pretty important. Signaling by solitons is promising as a future technology in trans-oceanic transmission. This was developed by Linn Mollenauer, a brilliant engineer at Bell Labs. It has not yet been put into practice, but it will be some day. The interesting thing about it is that classical signal theory is entirely linear, and the main point of soliton signal transmission is that the equations are nonlinear. There have been experiments using solitons for high-speed communication in optical fibers. The digital signal is coded using "ones" and "zeros", and we can let "ones" be represented by solitons. A key property of solitons is that they are exceptionally stable over very long distances. This offers the potential of considerably higher capacity in communication using optical fibers. Soliton theory also finds applications in quantum field theory and solid state physics, and in modeling biological systems.

"As for the theoretic importance: the KdV equation is completely integrable, one of an astonishing number of other completely integrable systems [that] were discovered. Completely integrable systems can really be solved in the sense that the general population uses the word solved. When a mathematician says he has solved the problem he means he knows the solution exists, that it's unique, but very often not much more.

"Now the question is: Are completely integrable systems exceptions to the behavior of solutions of non-integrable systems, or is it that other systems have similar behavior, only we are unable to analyze it? And here our guide might well be the Kolmogorov-Arnold-Moser theorem which says that a system near a completely integrable system behaves as if it were completely integrable. Now, what 'near' means is one thing when you prove theorems, another when you do experiments. It's another aspect of numerical experimentation revealing things. So I do think that studying completely integrable systems will give a clue to the behavior of more general systems as well. Who could have guessed in 1965 that completely integrable systems would become so important?"

In tribute to the soliton, Peter Lax has even written a haiku, a three-line poem.

"I got this idea from an article by Marshall Stone where he wrote that the mathematical language is enormously concentrated, like haikus. I thought I would take it one step further and actually express a mathematical idea by a haiku.

"Speed depends on size
Balanced by dispersion
Oh, solitary splendor."

Scattering theory

In the late 1950s Ralph Phillips of Stanford University approached Peter with an unexpected proposition: to do something with scattering theory. Peter described this collaboration as "One of the great pleasures of my life!", adding that "Ralph Phillips is one of the great analysts of our time and we formed a very close friendship." Phillips was then already the very well-known co-author, with Einar Hille, of a monumental treatise on functional analysis and semigroups.

Lax: Phillips thought I was lazy. He was a product of the Depression, which imposed a strict discipline on people. He thought I did not work hard enough, but I think I did!

The work of Lax and Phillips was expounded in two books, *Scattering Theory* and *Scattering Theory for Automorphic Functions*. Years later, when Peter wrote his survey *Hyperbolic Partial Differential Equations*, he included a chapter on scattering. At that time, he found his own earlier books with Phillips to be hard

	DIRECT	**INDIRECT**
GIVEN	OBSTACLE'S SHAPE and PROPERTIES	SCATTERING DATA
UNKNOWN	SCATTERING DATA	OBSTACLE'S SHAPE and PROPERTIES

reading. In order to write the chapter for *Hyperbolic Partial Differential Equations*, he made a fresh start and produced a more accessible exposition, which he now recommends to readers. He feels that he and Phillips made a mistake writing their expository books at a time when they were still in the midst of the research. He says it is better to wait a few years and then do the expository work.

So then, what is scattering theory?

I remember once standing with my son, Daniel, on the Rio Grande Gorge Bridge near Taos, New Mexico. We dropped a stone and watched it drop down far below into the river. Seconds later, we heard it splash in the water. Knowing roughly the speed of sound, we were able to estimate the height of the bridge.

This is a simple "scattering experiment". We sent a signal (the stone) into an obstacle (the river) and received a signal back (the sound wave). From the return signal we were able to deduce some information (its distance) about the obstacle. The sound of the splash is an example of "scattering data". The mathematical transformation, from the speed and mass of the rock entering the water to the speed of the sound wave coming back from the water, is an example of a "scattering operator".

More sophisticated scattering experiments are done every day in physics laboratories like CERN. There physicists shoot streams of particles into targets, where they interact with atomic nuclei. These interactions generate outcoming particles, which physicists observe. They use these observations to try to understand the nuclear structures where they were generated.

Scattering theory as a major method of analyzing quantum mechanical interactions is largely credited to Werner Heisenberg. The main experimental technique in quantum theory is shooting beams of particles at atomic nuclei. The collision of the incoming particles with the nucleus can result in many different outcomes, and this is the main source of information about the structure of the nucleus. The interaction of the incoming particle with the nucleus takes place over a very short time interval and at very short distances, which are quite inaccessible to direct observation.

The experimenter knows what signal he sent in and what debris he sees coming out. Heisenberg proposed that the "scattering operator", the transformation from the incoming beam to the observed scattering data, should be the main mathematical representation of the experimental outcome, since it contains all the available experimental information.

In the time-scale of such an experiment, the experimenter observes the wave very long before and very long after the interaction—at time "minus infinity" and time "plus infinity", so to speak. The observable properties of the object being struck by the waves are represented by the scattering operator, the operator that relates the incoming waves at time minus infinity to the outgoing waves at time plus infinity. This "scattering theory" became a central topic in quantum mechanics.

Lax and Phillips saw how the scattering approach could be applied to sound and light propagation. When sound waves or light rays are reflected and absorbed by an obstacle, the problem is again, How does the obstacle interact with and affect the incoming waves or rays? This classical problem is called "the direct problem" of scattering. One knows the shape of the obstacle and its "boundary conditions". Given the state of the exterior of the obstacle at time $t = 0$, one wishes to predict the future motion there. In quantum mechanics, on the other hand, the results of the scattering are observed. What is unknown, what is being sought, are the properties of the obstacle (the nucleus of an atom). This is called "the inverse problem of scattering". One seeks to analyze the nucleus of the atom (the "obstacle") by how it "scatters" the rays.

The classical direct problem is also called an "exterior mixed initial-boundary value problem", since one is given both the boundary conditions and the initial state in the exterior. (In fact, my own 1962 dissertation under Peter Lax was about mixed initial-boundary problems for general hyperbolic systems with constant coefficients.) Phillips's suggestion to Peter amounted to looking at this classical exterior problem from the viewpoint of scattering theory. One would interpret the effect of the obstacle as a perturbation of the evolution in free space, from time $t = -$ infinity to time $t = +$ infinity.

Phillips's superb familiarity with linear operator theory made it natural for him to seek a connection with the scattering operator. Peter eagerly took up his challenge, and they spent many fruitful years developing this theory in surprising and productive ways.

The inverse problem is also of interest in classical wave propagation. In geophysical exploration, direct observation of an obstacle may be impossible, and scattering data are used to infer its properties. Oil exploration uses signals generated by detonations that are propagated through the earth and through the oil reservoir and are recorded at distant stations.

(Peter said, "If you know the distribution of the densities of materials and the associated wave speeds, then you can calculate how signals propagate. That's the direct problem. The inverse problem is that if you know how signals propagate, then you want to deduce the distribution of the materials. Since the signals are discontinuities, you need the theory of propagation of discontinuities. It's somewhat similar to the medical imaging problem, also an inverse problem. Here the signals do not go through the earth but through the human body. You have to understand the direct problem very well before you can tackle the inverse problem.")

In three-dimensional wave propagation, either with or without an obstacle, an initial disturbance with a finite domain of existence eventually moves further and further away from the obstacle. As more and more time passes, the disturbance moves out to infinity, so to speak. It becomes zero in every bounded region. This is sometimes called "Huygens principle". Moreover, since the wave equation is symmetric in time, the same conclusion holds if we look *backward* in time. In the very remote past, the disturbance would have been felt only in very remote regions. When the disturbance interacts with an obstacle, its motion is "perturbed" from the evolution generated by its remote past and sent to a different remote future state. But "very remote regions" and "remote past and future" are not precise mathematical terms!

Lax and Phillips found a way to give a mathematically meaningful definition of the desired scattering operator. Given the perturbed motion $u(t)$, they associate to both its remote past and its remote future certain motions in free space (without obstacles). In fact, there is a unique free-space wave motion or vibration, $u-$, to which the perturbed motion converges as we go backwards in time, and there is another one, $u+$, to which it converges as we go forward in time. They can therefore define a scattering operator associated to the obstacle: it is the transformation from $u-$ to $u+$. This scattering operator makes precise the notion of a perturbation of a motion originating far away, in the remote past, into a different outcome, again far away, in the remote future.

In Peter's words, "We had a new way of viewing the scattering process with incoming and outgoing subspaces. We were, so to say, carving a semi-group out of the unitary group, whose infinitesimal generator contained almost all the information about the scattering process. We applied that to classical scattering of sound waves and electromagnetic waves by potentials and obstacles."

All this takes place in the context of Hilbert space, which is the standard mathematical context for quantum mechanics and which Kurt Friedrichs had developed as the context also for classical wave propagation. Hilbert space is a natural generalization of classical Euclidean space of two and three dimensions. The solution of the wave equation at any time t, the displacement from equilibrium of the motion or vibration, is a vector in that Hilbert space, with a norm or magnitude defined as its "energy". This "energy" is the sum of the integrals of the squares of the first derivatives of the solution. The derivative with respect to time t is the velocity, of course, and the integral of its square corresponds to kinetic energy. The derivatives with respect to the space variables correspond to potential energy. The magnitude of the motion, the total energy, is the sum of the kinetic and the potential energy.

The Lax-Phillips scattering theory is a chapter in the theory of linear transformations of Hilbert space. It brings in a whole collection of new ideas which are physically motivated and are expressed geometrically in this infinite-dimensional function space. The resulting theory is a playground for many aspects of functional analysis, including groups of operators and analytic function theory.

If the obstacle is convex or star-shaped, so that rays cannot be trapped by indentations in the obstacle, Lax and Phillips show that energy is propagated to remote distances at a rate proportional to the square root of time elapsed. The geometric function-space flavor of the mathematical structure is fully exploited.

A huge unexpected bonus sprang from a suggestion by the very famous Russian mathematician, Israel Moiseyevich Gelfand, to consider Lax-Phillips scattering in non-Euclidean space. Following Gelfand's suggestion, the Russian investigators Boris Faddeev and Ivan Pavlov connected the scattering results to classical questions in number theory involving the so-called Eisenstein series. In the space of functions in the hyperbolic plane invariant under the modular group, the Eisenstein series are the eigenfunctions corresponding to the continuous spectrum of the Laplace-Beltrami operator.

On this topic, Peter wrote: "Following a very interesting discovery of Faddeev and Pavlov, we studied the spectral theory of automorphic functions. We elaborated it further, and we had a brand new approach to Eisenstein series for instance, getting at spectral representation via translation representation. And we were even able to contemplate—following Faddeev and Pavlov—the Riemann hypothesis peeking around the corner."

If a certain estimate regarding the scattering theory in non-Euclidean space could be proved, it would imply the Riemann hypothesis. Unfortunately, it seems that proving the needed estimate is no easier than the Riemann hypothesis itself.

"Whether this approach will lead to the proof of the Riemann hypothesis, by stating it purely in terms of decaying signals and cutting out all standing waves, is unlikely. The Riemann hypothesis is a very elusive thing. You may remember in *Peer Gynt* there is a mystical character, the Boyg, which bars Peer Gynt's way wherever he goes. The Riemann hypothesis resembles the Boyg!"

Differentiability of the Pólya function

"I knew Pólya quite well," writes Peter, "having taken a summer course with him in 1946. The differentiability question came about this way: I was teaching a course on real variables, and I presented Pólya's example of an area-filling curve, and I gave as homework to the students the problem of proving that it's nowhere differentiable. Nobody did the homework, so then I sat down and I found out that the situation was more complicated."

First one must notice the seeming paradox: an "area-filling curve". A familiar curve like a circle or an ellipse "obviously" takes up zero area. But area-filling curves have been known for a century. There is an example by David Hilbert which passes through every point of the unit square. Of course one does not try to draw Hilbert's curve with a pencil; it's defined as the limit of a sequence of approximating curves.

Pólya's example fills a triangle instead of a square. That is to say, he defines a continuous mapping of the unit segment of the real axis *onto* the interior of a triangle T. The result, being the continuous image of a line segment, is by definition a continuous curve. T is assumed to be a right triangle, not isosceles.

Let x be any point on the unit segment; write it in binary expansion as $x = .d_1 d_2 \ldots$, the d_i zeros or ones. To specify the image $P(x)$ in the interior of T, first draw the altitude onto the hypotenuse, dividing T into two similar subtriangles. Since T is not isosceles, the two subtriangles are unequal. Call the smaller one T_s, the larger one T_l. Define T_1 to be T_s if the first digit d_1 of x is 0, T_l if d_1 is 1. $T_2, T_3, \ldots$ are defined recursively, with T_{n-1} taking the place of T and d_n replacing d_1. The sequence of triangles $T_n(x)$ is nested, and their diameters converge to zero, so the sequence $T_n(x)$ has exactly one point in common, its limit point. That point is defined to be $P(x)$, the image of x.

Every point inside the triangle is covered, for every point is the intersection of a unique sequence of nested triangles with successive choices of left and right, which come from a sequence of 0's and 1's, which is to say, from a point on the unit interval. Some points of the unit interval, for example the midpoint, can be written in two different ways in binary expansion, as .1000... and as .0111... for example. It is easy to see that the mapping T assigns the same point to both expansions, the vertex of the right triangle. It is not difficult to see that the mapping is continuous (points that are sufficiently close on the unit interval will map onto points that are arbitrarily close in the interior of the triangle).

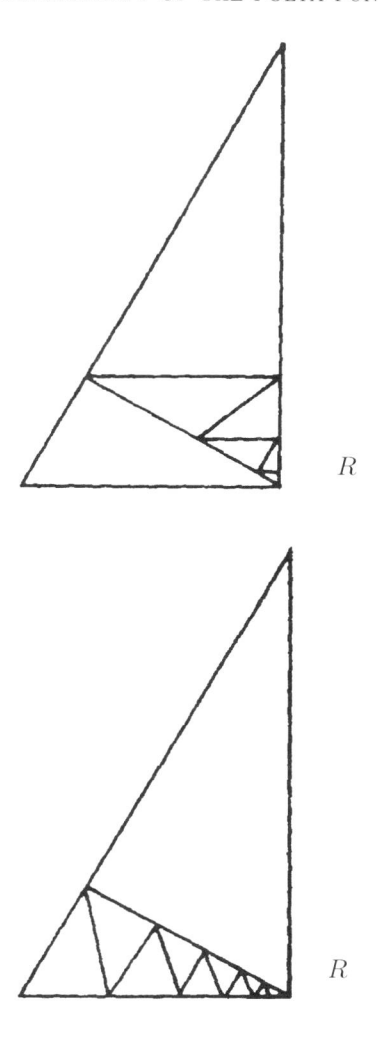

This curve is ill-behaved—it actually fills up a whole triangle!—so it was natural to expect that it would be nondifferentiable and that students could prove it. When they couldn't, Peter sat down with the problem and discovered quite a surprise. The curve might be differentiable nowhere or it might be differentiable almost everywhere or it might be nondifferentiable *almost everywhere* while still differentiable on an uncountable set! The details of this distinction depend on the shape of the triangle. A spoilsport might say, "So what? Who cares? What use is this?" Indeed, it may not be very useful. No matter. If you have a mathematical soul, you will find this result astonishing, or at least amusing.

DIFFERENTIABILITY THEOREM 1. *Denote by A the smaller angle of T.*

(a) *If A is strictly between* 30 *degrees and* 45 *degrees, P is nowhere differentiable.*

(b) *If A is strictly between* 15 *degrees and* 30 *degrees, P is not differentiable on a set of measure* 1; *it has derivative zero on a nondenumerable set.*

(c) *If A <* 15 *degrees, P is differentiable on a set of measure* 1*, and* $P'(x) = 0$ *there.*

How did Peter prove it? By straightforward elementary calculation, using the definition of the derivative! It was published in *Advances in Mathematics* (vol. 10, 456–464) in 1973 and dedicated to Garrett Birkhoff.

Lax-Milgram lemma

Arthur Milgram was a topologist, a good friend of Lipman Bers, who decided to learn some analysis by spending a year at the Courant Institute as a visiting researcher. He asked Peter Lax to suggest a possible research topic.

Peter told him about the variational approach to solving self-adjoint boundary-value problems, which yields a weak solution by means of a classical theorem of functional analysis, the Riesz representation theorem. Peter suggested trying to generalize it to non-self-adjoint problems.

The Riesz representation theorem says that any bounded linear functional on a Hilbert space H is just an inner product with some vector in the space. (The inner product, of course, is a symmetric function of two vectors.) Milgram solved the non-self-adjoint problems by generalizing the inner product to a possibly asymmetric bilinear function of two vectors.

"If $B(x, y)$ is a bounded function of pairs of vectors in H, and $B(x, y)$ is linear with respect to x, for fixed y, and $B(x, y)$ is skew-linear with respect to y, for fixed x, and $B(y, y)$ is bounded from below by some constant times the norm squared of y, then EVERY bounded linear functional on H is equal to $B(x, y)$, for some unique vector y."

This theorem became a standard tool for non-self-adjoint elliptic boundary value problems using coercive inequalities. "My contribution was only to propose it," says Peter, "but I don't know how to get my name off it."

Algebraic hyperbolicity

In studying hyperbolic partial differential equations with constant coefficients, one can consider either a single differential equation of degree n or a system of n first-order equations, written as a single matrix equation. It is plausible but not completely clear that these two kinds of problems are equivalent. In case there are three independent variables (one time variable, two space variables), Peter noticed that there are the same number of free parameters for the set of *symmetric n-by-n* matrices as for the set of n-th order single equations. After a standard change of variables, the number of free parameters in a symmetric system of n first-order partial differential equations is the same as in a single equation of order n:

$$n + (n(n+1))/2.$$

So he conjectured that each of those higher-order equations was just the determinant of some first-order system:

"Every homogeneous hyperbolic polynomial of degree n in 3 variables, $P(a, b, c)$, can be written as a determinant: $P(a, b, c) = \det(aI + bB + cC)$, where I is the

identity matrix, B a diagonal matrix, and C an n-by-n real, symmetric matrix."
"Hyperbolic" means that for any real values of b and c, the equation $P(a,b,c) = 0$ has only real solutions a.

Why was this conjecture plausible? Because the determinant of a real symmetric matrix has only *real eigenvalues*. And the number of free parameters matches in both the first-order matrix equation and the single higher-order equation. Moreover, one easily checks that the conjecture is true when $n = 2$.

In 1960, when I was ready to start dissertation research, Peter showed me his list of possible thesis topics. I had little grasp of most of the problems on the list, but this one looked elementary, so I chose it.

I got stuck trying to prove the case $n = 3$. Peter suggested I consult with André Weil, a preeminent algebraic geometer who sometimes came up to New York from Princeton. I found Weil at teatime in the Courant Institute lounge on the thirteenth floor and approached him cautiously. He answered tersely: "Such questions involving real roots are very difficult."

I ultimately wrote a dissertation on a completely different topic.

Nearly fifty years later, when I was asked to write a review of Peter's *Selected Papers*, I read, on page 587 of Volume 2, "Recently A. S. Lewis et al. succeeded in deducing this conjecture from an observation of Helton and Vinnikov, based on a deep result of Vinnikov."

Vinnikov's cited paper had been published in 1993 and was twenty-six pages long.

Epilogue

A few months ago I was in New York, looking into NYU's archives. Peter, Louis Nirenberg, and Cathleen Morawetz are the only people still living in New York who were my grad school teachers. (Joe Keller now lives in Palo Alto, California.) Although Peter, Louis, and Cathleen are close friends, they don't see each other very often. I had the notion that if I got the three of them together, reminiscences would flow and I could take copious notes. They did all agree to meet at the Courant Institute at teatime on a Friday afternoon. Cathleen couldn't make it, but Joe Keller was there, on a visit to New York. Everybody sipped their tea, and nobody reminisced. I asked Joe what he was doing these days, and he said he was goofing off. Then he admitted he was looking over some old unpublished papers, trying to dig out the mistakes. I advised him to publish them as is and let others find the mistakes. I am sure he will ignore my advice.

While I was working on this book, Peter visited New Mexico for a few days. We were guests at a dinner party given by Porter Dean Gerber, one of Peter's Ph.D. students, who has retired to his boyhood home of Santa Fe after a career at IBM Yorktown Heights. Peter gave a talk on multiple eigenvalue phenomena to the University of New Mexico math department in Albuquerque. I was afraid it would be poorly attended, because it was between semesters, but the room was full and the talk was a success. We had lunch with his student and collaborator, Burt Wendroff. Burt lives in Los Alamos, and we met halfway between Santa Fe and Los Alamos, at "Buffalo Thunder", an incredibly ornate and pretentious gambling resort belonging to a very small Indian tribe, the Pojoaque. Peter and Burt reminisced while I taped their conversation. Then Peter and I enjoyed a drive through the Jemez Mountains.

I knew that while Peter was in New Mexico he would want to visit Francoise Ulam, the widow of his old friend, the famous "father of the H-bomb", Stan Ulam. I went to see her at the retirement home where she was living. She took a long time to answer my knock, and when she came to the door, she was angry that I interrupted while she was "gathering her thoughts." But her expression changed when I explained that I was there to let her know that Peter Lax was planning to come to New Mexico and would call on her. "I will be very happy to see Peter Lax," she declared. Peter called her on the phone from New York. But she died before he could see her again.

The appendices that follow are documents that would have interrupted the main text. Appendix 1 is an account of Anneli Lax's academic and scientific career. Appendices 2 and 3 are Peter Lax's biographical articles about John von Neumann and Richard Courant, respectively. Appendix 4 contains Peter's Curriculum Vitae and list of publications, followed by his statement and proof of the Closed Graph Theorem (needed for the Lax-Richtmyer Equivalence Theorem) in Appendix 5. Appendix 6 is a list of Peter's Ph.D. students, copied from the Mathematics Genealogy Project. Appendix 7 is a memoir of John Lax by his mentor, Abbott Gleason, followed by John Lax's article about Chicago jazz in Appendix 8.

APPENDIX 1

Anneli Lax

To introduce Anneli's scientific and academic career, I cannot do better than to quote the obituary by Mark Saul from the AMS *Notices*, August 2000.

"Anneli's life was filled with service and friendship to the mathematical community. Lax was a gifted mathematician, a master of language, a remarkable teacher. Yet the defining characteristic of her life was perhaps not any of these gifts, but rather how she used them. She seemed to have an inner drive to share with others what she could do, and this drive led her from one endeavor to another in the service of mathematics and mathematicians."

Anneli earned a bachelor's degree from Adelphi University in 1942 and moved on to graduate work at New York University. She took the graduate course in complex variables in the spring of 1943, and there she met Peter Lax, who was still an undergraduate.

She found that Adelphi had not prepared her well for graduate work in math, but she discovered a mistake in a book Richard Courant had assigned her to read, and Courant was deeply impressed. So he recruited her to help translate into English his book *Methods of Mathematical Physics*. In 1992 she recalled that Courant had often asked her to edit things other people had written. "In fact, he claimed that he hired me because I seemed more literate than most people...I ended up doing everything. I even made page dummies. That was kind of fun: it was like playing with paper dolls." She was modest about her editorial skills. "I have to understand every darn little step, which slows me down terribly and which is one of the reasons that I never learned very much, but it is good for checking errors and making sure that everything is okay."

Ultimately, she attained great recognition as a book editor for the Mathematical Association of America.

Anneli received her Ph.D. in 1956 with a thesis under the supervision of Richard Courant. The title was "On Cauchy's problem for partial differential equations with multiple characteristics". She proved that a sufficient condition which E. E. Levi had given for differential equations to be hyperbolic is also necessary. It was published in *Communications on Pure and Applied Mathematics* in 1956. She rose through the faculty ranks to become a professor at NYU in 1961.

"I started teaching at NYU in the mid-forties, before I had the degree.... In all of the many years I've taught, I now, in retrospect, think that I didn't really understand teaching until the last ten years or so." To help NYU freshmen who had difficulty in learning mathematics, she designed and helped teach a course in mathematics and writing. She worked closely with high school teachers to make mathematics more attractive to New York children. A particular interest of Anneli's was the relation between mathematics and language, and she proposed new ways to connect mathematics teaching with language teaching. She had a knack

Anneli Lax teaching at Courant Institute, 1979

for examples that helped high school teachers understand advanced mathematical concepts.

John Devine, an NYU professor of education whom she worked with, recalled, "We brought the math teachers and English teachers together for joint sessions after school. This was unheard of. They didn't know each other. Anneli ran these sessions like a mathematical psychoanalyst. She was able to get the English teachers to lose their fear of introducing mathematical terms and concepts and procedures into their English classes. On the mathematics side, she was able to get the math teachers less afraid of word problems." Soon she started tutoring students in inner-city high schools. "Anneli would come into the tutoring rooms and work with the kids themselves. This was beautiful to behold. She would sit at a tutoring table with some ninth-grade girl who had poor reading and writing skills. Anneli would always begin where the student was. Her interest was in knowing the student's thought processes. She would do everything she could to try to get at the way kids were thinking, not the way she herself was thinking. She would get them talking, and suddenly the kid would be saying, 'I went to the store this morning and helped Grandma figure out her food stamp budget.' So Anneli would become interested in the food stamp budget. She would get around to the textbook, but only after understanding the kid's view."

Anneli wrote, "My concern has been access to mathematics, and my efforts have been directed to making sure that our schools do not deprive students of learning how to think for themselves by developing, among other skills, one of their natural talents: looking at the world mathematically...I have been trying to promote attention to use of language in all learning, particularly learning mathematics, and developing the art of listening and reading so that we can apply this art to looking at our students' emerging ideas.... Let us practice what we preach: read and write carefully, avoid trendy slogans, and go beyond mathematical correctness, syntactic correctness, and political correctness in serving our discipline in our individual ways."

In 1957 Courant recommended her for editor of the New Mathematical Library (NML). The New Mathematical Library sought to make accessible to high school students, and to the general public, deep results in mathematics, reported by research mathematicians, without sacrificing technical accuracy. Under her, it became one of the foremost expository series in mathematics. Her meticulous care was devoted to making every book in the series nearly perfect. She recounted that in editing one book in that series, *An Introduction to Inequalities* by Beckenbach and Bellman, "I wrote the last chapter and inserted it into the manuscript. Each of them thought that the other had written it and never said boo."

For thirty-three years, she took care of the NML's acquisition, copy and mathematical editing, cover design, and typesetting. The New Mathematical Library grew to more than forty volumes. Her work was celebrated by the mathematics profession. When she died, the series was named in her honor.

In addition, for many years she volunteered in recording books for the blind at a studio in New York. She read aloud books such as Dunford and Schwartz's *Linear Operators* and Granville, Smith, and Longley's *Calculus*.

In 1992 she retired. After that, she had time to hike, swim, ski, and canoe at Loon Lake. In 1995 the MAA gave her the Yueh-Gin Gung and Dr. Charles Y. Hu Award for Distinguished Service, with the words, "No other person in the history of the Association's book publishing effort has played a larger role in developing and nurturing a book series." The February 1995 issue of the *American Mathematical Monthly*, carrying the announcement of that award, had as its front cover the following photograph of Anneli. The next time I saw her, I told her how beautiful it was. She answered, "You should have seen me ten years ago."

Anneli Lax (1922-1999)

Photograph by Henry Lax, used on the cover of the American Mathematical Monthly, *February 1995. Courtesy of James Lax.*

In 1997 Mount Holyoke College awarded her an honorary doctorate. Also in 1997, her son Jimmy diagnosed her with cancer of the pancreas. This cancer usually is rapidly fatal, but she survived for two years. On September 24, 1999, she passed away at home in New York.

Elena Marchisotto, a professor at California State University, Northridge, wrote her doctoral thesis under Anneli's direction. Elena remembered Anneli this way: "One evening at Loon Lake Anneli suggested we go to pick some berries. I had visions of strolling down a lane with a white basket daintily selecting delicate fruit. Anneli, however, drove us deep into the woods, walked us down bug-infested paths to berry bushes. In order to retrieve the fruit we had to be fierce with the prickly growths. Anneli dove right in and filled her bucket to the brim! I loved Anneli's quiet determination, her openness and acceptance of the weaknesses of others and her joy in simple pleasures. She is sorely missed."

APPENDIX 2

John von Neumann: The early years, the years at Los Alamos, and the road to computing

Peter Lax
March 1, 2005

Today, nearly 50 years after his death, he looms larger than ever as the prophet of the age of technology. His genius was in his mathematics, and a mathematical way, coupled to uncommon common sense, pervaded his thinking about everything. The story starts with the birth of the hero, on December 28, 1903, in Budapest.

Von Neumann, at about age 6. Photo from the collection of Marina von Neumann Whitman.

He was the oldest of three boys of an upper-middle-class Jewish family; his father Max was a banker. The end of the 19th and the beginning of the 20th century were heady times for Budapest.

The school system, reformed by von Kármán's father, was sensitive to unusual talent, so it is not surprising that Rácz László, teacher of mathematics at the Evangelical Gymnasium (among whose students more than 50% were Jewish) recognized immediately the Neumann boy's extraordinary gift. He informed Jancsi's parents, and also József Kürschák, the Nestor of the Hungarian mathematical community, and it was arranged that the young Neumann boy would receive special instruction. His first tutor was Gábor Szegö, himself a former prodigy, later a professor at Königsberg and then at Stanford; Mrs. Szegö liked to recall that her husband came home with tears in his eyes after his first encounter with the young boy genius. When Szegö left for Germany, Michael Fekete, later of the Hebrew University in Jerusalem, took over as tutor. Von Neumann's first publication was a joint paper with Fekete on the transfinite diameter, written in 1922, when von Neumann was 19; Fekete devoted the rest of his long scientific career to this subject.

Most mathematicians shy away from mathematical problems posed in a nonmathematical setting. Not for all, to be sure; but few have embraced real-world problems so wholeheartedly as von Neumann. According to his closest friend, the mathematician Stan Ulam, von Neumann's thinking was not geometric, nor tactile, but of an algebraic sort: games played with algebraic symbols on the one hand, and an interpretation of their meaning on the other. This may explain his ability to think in so many different milieus.

After young John finished the gymnasium, his father decided that mathematics was an impractical career; chemical engineering was a much more promising profession. So he went off, first to Berlin,* and two years later to Zürich. But he also enrolled at the University of Budapest for a PhD in mathematics, to be earned mostly in absentia.

In Berlin von Neumann began to write his mathematical dissertation on a technical-sounding but philosophically deep subject, "The Introduction of Transfinite Ordinals." It was eventually published under the title "An Axiomatisation of Set Theory." Its purpose was to resolve a slowly brewing crisis in mathematics.

This crisis split the mathematical community into two camps: the intuitionists, who would severely circumscribe the way infinite sets are handled, and the formalists, who believed that proper axiomatization, in the spirit of Euclid, would free us to handle infinite sets to our heart's desire, and at the same time deliver us from contradictions. The leader of the formalists was David Hilbert in Göttingen.

In 1923 von Neumann went to Zürich to begin his studies of chemical engineering. There he made the acquaintance, or rather they of him, of two leading mathematicians, George Pólya and Hermann Weyl, the latter one of the leaders of the intuitionists. In 1926 he received his degree in Zürich, and shortly afterward his degree in mathematics in Budapest. He was not yet 23 years old.

His work on the foundations of set theory attracted the attention of the aging Hilbert in Göttingen, and his growing reputation earned him a grant from the Rockefeller Foundation to spend a year at Göttingen. When he arrived there he found that the burning issue of the day was not the foundation of set theory, but

---

*When he moved to Germany, he started to use the appellation von, a loose translation of the title Margittai awarded to his father.

the newly created quantum mechanics. The mathematics needed to clear up these new theories of Heisenberg and Schrödinger occupied von Neumann, on and off, for the rest of his life. The theory he created of unbounded self-adjoint operators in Hilbert space gives a logically satisfactory basis for quantum mechanics, and it is a basic building stone of modern mathematics as well. Furthermore—and this was typical of von Neumann—he not only laid the foundations but showed how they could be applied in specific, physically interesting situations.

By this time von Neumann's reputation was soaring. He was appointed Privat Dozent in Berlin, and subsequently in Hamburg; he was invited to lecture all over Europe. But by the end of the 1920s his eye was on America, partly because of the paucity of jobs in Europe, but also because he profoundly distrusted and feared the political instability of Europe, which he saw coming long before most other people did. So when an invitation from Princeton arrived in 1929 to lecture on mathematical physics, mainly the new quantum mechanics, he accepted with alacrity. For the next four years he divided his time equally between Princeton and Germany.

The idyllic 50/50 arrangement come to an abrupt end in 1933 for two reasons: Hitler's rise to power, and von Neumann's appointment as professor at the newly created Institute for Advanced Study, also at Princeton. This was a very prestigious position; Albert Einstein and Hermann Weyl were fellow professors, and Gödel was to join later.

The mid-1930s were a fertile period for von Neumann. In collaborations with Francis Murray, he made one of his most enduring discoveries, a theory of rings of operators, today called von Neumann algebras. At the same time the gathering political crises convinced him that war was inevitable, and that it would come soon. He also foresaw that it would lead to the destruction of the European Jews, much as the Armenians had been destroyed by the Turkish government during the First World War.

It is therefore not surprising that, feeling keenly that war was coming, he thought of ways in which he might use his mathematical talents to help America prepare for war. The most mathematical part of warfare at that time was ballistics. The Aberdeen Proving Grounds were conveniently located not far from Princeton, and so he threw himself energetically into the study of explosions and shock waves. In the process he almost became an army lieutenant in the Department of Ordnance, except that he was (barely) over the age limit of 35, and the Secretary of War would make no exception. Soon, his fame as a practical applied mathematician began to spread, just as his fame as a brilliant pure mathematician had spread fifteen years earlier.. Early in 1943 he was sent over to England to help out in antisubmarine and aerial warfare; he was able to help and in turn learned a great deal from the British about detonations. Soon he was able to put all his recently acquired knowledge to use for the most important project of the war, the making of an atomic—more precisely, nuclear—bomb.

When von Neumann arrived at Los Alamos, there were many open problems, each of which had to be overcome for the successful construction of a plutonium bomb. An isotope of plutonium fissions spontaneously and emits neutrons, in sufficient quantity to predetonate any bomb unless it has been assembled fast enough. Implosion was the method most promising for assembly. Von Neumann's earlier knowledge of high explosives steered him to a safe and fast way to accomplish it.

This and his many other technical contributions to solving physics and engineering problems established his reputation as the person to consult. He was admired by the brightest stars at Los Alamos: Oppenheimer, Bethe, Feynman, Peierls, Teller, and many others; they acknowledged him as their superior for sheer brain power.

Nuclear weapons cannot be designed by trial and error; each proposed design has to be tested theoretically. This requires solving the equations of compressible flows, governed by nonlinear equations.

Von Neumann came to the conclusion that analytical methods were inadequate for the task, and that the only way to deal with equations of continuum mechanics is to discretize them and solve the resulting system of equations numerically. The tools needed to carry out such calculations effectively are high-speed, programmable electronic computers, large-capacity storage devices, programming languages, a theory of stable discretization of differential equations, and a variety of algorithms for rapidly solving the discretized equations. It is to these tasks that von Neumann devoted a large part of his energies after the war. He was keenly aware that computational methodogy is crucial not only for designing weapons, but also for an enormous variety of scientific and engineering problems; understanding the weather and climate particularly intrigued him. Computing, he realized, can do more than grind out by brute force the answer to a concrete question.

In a talk delivered in Montreal in 1945, when fast electronic computers were merely figments of his imagination, he said:

"We could, of course, continue to mention still other examples to justify our contention that many branches of both pure and applied mathematics are in great need of computing instruments to break the present stalemate created by the failure of the purely analytical approach to nonlinear problems. Instead we conclude by remarking that really efficient high-speed computing devices may, in the field of nonlinear partial differential equations as well as in many other fields which are now difficult or entirely denied of access, provide us with those heuristic hints which are needed in all parts of mathematics for genuine progress. In the specific case of fluid dynamics these hints have not been forthcoming for the last two generations from the pure intuition of mathematicians, although a great deal of first-class mathematical effort has been expended in attempts to break the deadlock in that field. To the extent to which such hints arose at all (and that was much less than one might desire), they originated in a type of physical experimentation which is really computing. We can now make computing so much more efficient, fast and flexible that it should be possible to use the new computers to supply the needed heuristic hints. This should ultimately lead to important analytic advances."

Everybody knows that John von Neumann was the founding father of the modern computer; not everybody realizes that he was also the founding father of computational fluid dynamics. One of von Neumann's fundamental contributions to the theory of difference equations was a notion of stability; an important test for it is named after him. As originally stated by him, this test implies the stability only of linear equations with constant coefficients; but von Neumann boldly asserted that it applies also to systems with variable coefficients—and so it turned out to be.

The deepest idea of von Neumann for computing compressible flows is shock capturing. This means that shocks and other discontinuities that inevitably arise in such flows are represented in the discrete approximations not as discontinuities but as rapid transitions, and all points in the flow field are treated as ordinary points. In a calculation performed in 1944, von Neumann studied the flow of gas in a tube one of whose ends is closed off; initially, the flow is directed toward the closed end. Subsequently a shock is formed and propagates backward, opposite to the flow direction. The paths of the particle change direction abruptly at the shock, because the velocity field, which they follow, changes discontinuously across the shock. Von Neumann observed that the particle paths are wobbly near the shock; this indicates that the velocity field is oscillating there. These oscillations are a consequence of the dispersive nature of the difference equations that von Neumann employed. Artificial viscosity, subsequently introduced in a paper with Richtmyer, eliminated the unwanted oscillations.

The tragedy of von Neumann's early death robbed mathematics and the sciences of a natural leader and an eloquent spokesman, and deprived a whole younger generation of beholding the most scintillating intellect of the 20th century.

Bibliography

J. von Neumann and M. Fekete, Über die Lage der Nullstellen gewisser Minimum-polynome, *Jahresbericht* **31** (1922), 125-138.

J. von Neumann, Eine axiomatisierung der Mengenlehre, *J. für Math.* **154** (1925), 219-240.

A. H. Taub, ed., Proposal and analysis of a numerical method for the treatment of hydrodynamical shock problem, in *John von Neumann Collected Works*, Vol. 6, No. 26, Macmillan, New York, 1963.

APPENDIX 3

The life of Richard Courant

By Peter Lax

DURING HIS LONG and adventurous life Courant achieved many things in mathematics: in research and the applications of research, in the exposition of mathematics and the education of students, and in administrative and organizational matters. To understand how he, essentially an outsider both in Germany and the United States, accomplished these things we have to examine his personality as well as his scientific works. But let's start at the beginning.

Courant was born on January 8, 1888, in the small town of Lublinitz in Upper Silesia, now part of Poland but then of Germany. His father, Siegmund, was an unsuccessful businessman. The family moved to Breslau, and soon the precocious Richard was beginning to support himself by tutoring. In the gymnasium he came under the influence of a charismatic teacher of mathematics, Maschke, who inspired specially selected, talented students with a love of mathematics. Six years after Courant the young Heinz Hopf entered the gymnasium in Breslau and came under the tutelage of Maschke, who trained his special pupils by posing challenging problems. Many years later Hopf recalled that he was able to solve most of them, but was stumped every once in a while. "Courant could solve it," said Maschke.

This no doubt was the first bond in the intimate friendship that developed later between Courant and Hopf.

After the gymnasium Courant was ready to attend university lectures on mathematics and physics at the University of Breslau. Because of the weakness of the physics faculty he gravitated toward mathematics. He spent a semester at the University of Zürich, but still dissatisfied, he set out in the fall of 1907 for what his Breslau friends, Otto Toeplitz and Ernst Hellinger, described as the Mecca of mathematics, Göttingen.

Not long after his arrival in Göttingen Courant was accepted as a member of the "in-group" of young mathematicians whose leader was Alfred Haar, and which included Toeplitz, Hellinger, and von Kármán but did not include Hermann Weyl. These brilliant young men were attracted to Göttingen by its stellar mathematics faculty: Klein, Hilbert, Minkowski,, Runge, Zermelo, the fluid dynamicist Prandtl, and the astrophysicist Schwarzschild. For young Courant the shining light was Hilbert, and it was his great good fortune that in 1908 Hilbert chose him to be his assistant.

The next phase in his career was the writing of a dissertation. It is illuminating to go back more than 50 years to the dissertation of Riemann. Riemann proved the existence of harmonic functions by a variational method called Dirichlet's principle, according to which a certain quadratic functional—the Dirichlet integral for functions of two variables—assumes its minimum value. Weierstrass challenged the validity of this proof, because the existence of a minimum cannot be taken for granted. Weierstrass even gave an example of a fourth order functional whose minimum is not assumed by any function. In view of the basic nature of Riemann's work, there was a feverish effort by the leading mathematicians to furnish an alternative proof. Poincaré did it with a method he called "balayage." Carl Neumann derived and then solved an integral equation; this work paved the way to Fredholm's famous study of general integral equations, which in turn was followed by Hilbert's and subsequently Frederic Riesz's analysis of the spectrum of compact operators. Hermann Amandus Schwarz supplied the missing proof by his famous "alternating method." One hundred years later, by one of those twists that are not infrequent in mathematics, the alternating method, now called "domain decomposition," turned out to be the most efficient numerical method for solving Riemann's problem and more general problems, when the calculations are performed by computers with many processors in parallel.

Hilbert's way of filling the gap was to supply the missing step in Riemann's argument, the existence of the minimizing function. Curiously, according to Haim Brezis and Felix Browder, Hilbert's own work is incomplete. The crucial idea for fixing Dirichlet's principle was supplied in 1906 by Beppo Levi.

Back to Courant. Hilbert suggested to him as a dissertation topic to use Dirichlet's principle to prove the existence of various classes of conformal maps. Courant succeeded, and was awarded his Ph.D. summa cum laude in 1910. The same topic served for his habilitation dissertation in 1912.

Dirichlet's principle remained a lodestar for Courant; he kept returning to it throughout his career. He was fascinated not only by its use in theory but also by the possibility of basing numerical calculations on it, as was done by the young physicist Walther Ritz.

Courant liked to spice his lectures with remarks about the personalities of scientists, to render them more human. Thus, in a talk in Kyoto in 1969, his last public lecture, he described the work of Walther Ritz and recalled that Ritz died young, at the age of 31, of tuberculosis, and that he refused to enter a sanitarium for fear that it would prevent him from completing his life's work. Then Courant added that Walther Ritz was a member of the Swiss family whose hotels all over the world made their name synonymous with luxury.

In 1912 Courant married Nelly Neumann, a fellow student from Breslau; the marriage lasted only four years. They were joined in Göttingen by Courant's favorite intellectual cousin from Breslau, Edith Stein, who became a student and later assistant to the philosopher Husserl. She attained martyrdom as a Jew and posthumous fame as a saint of the Roman Catholic Church, canonized by Pope John Paul II in 1998.

Harald Bohr turned up in Göttingen in 1912; he and Courant became fast friends. They wrote a joint paper on the distribution of the values of the Riemann zeta function along the lines Re zeta = constant in the critical strip. This was Courant's only venture into number theory. The friendship with Harald Bohr later came to include Harald's brother, Niels, and lasted until the end of Niels's life.

The idyllic Göttingen life was shattered, like everything else in Europe, by the outbreak of the First World War; the flower of European youth was led to slaughter. Courant was drafted into the army; he fought on the western front and was seriously wounded. While in the trenches, Courant had seen the need for reliable means of communication, and came to the idea of a telegraph that would use Earth as a conductor. He consulted Telefunken, the German telephone company, and his teacher Carl Runge in Göttingen, who brought Peter Debye and Paul Scherrer into the project. In the end the Earth telegraph became a resounding success; equally important, the experience taught Courant how to deal successfully with people of all classes: officers of rank high and low, engineers, industrialists.

Courant's absence in the army did not make his and Nelly's hearts grow fonder. On the contrary, it made both of them realize their incompatibility. After their divorce in 1916 he found himself drawn to Nina Runge, daughter of Carl Runge, professor of applied mathematics in Göttingen, and she to him. They were married in 1919. They had much in common-a passionate love of music-but in many respects they were very different. Their marriage was a successfully shared life. They had four children: two boys, who became physicists, and two girls, a biologist and a musician.

The years 1918-20 were banner years for Courant. He proved that among all plane domains with prescribed perimeter, the circle had the lowest fundamental frequency. This was followed by a max-min principle that enabled him to determine the asymptotic distribution of eigenvalues of the Laplace operator over any domain, a result of great physical interest, established previously by Weyl with the aid of a min-max principle. Weyl's method leads naturally to upper bounds for the eigenvalues, Courant's to lower bounds. The combination of the two methods is particularly effective.

It was during this period that Courant's friendship with the publisher Ferdinand Springer matured. Courant encouraged Springer to enlarge his offering in mathematics. This led to Springer's taking over the *Mathematischen Annalen* and starting the *Mathematischen Zeitschrift*. Equally important was the new book series *Grundlehren*, affectionately known as the Yellow Peril for its yellow cover.

After the war Courant was offered and accepted a professorship in Münster, but this was merely a stepping stone for a position the following year in Göttingen, pushed through by Hilbert and Klein. The latter saw Courant—correctly—as one who would share his vision of the relation of mathematics to science, who would seek a balance between research and education, and who would have the administrative energy and savvy to push his mission to fruition.

The early 1920s were a tough time in Germany. The defeat in the First World War had demoralized large segments of society and had led to rampant inflation. Courant showed his resourcefulness by keeping things afloat, partly with the help of the far-sighted industrialist Carl Still.

In 1922 Courant's first book appeared, Hurwitz-Courant on *Function Theory*, the third volume in the Yellow Peril series. The first part, based on lecture notes of Hurwitz, was written from the Weierstrass point of view; its main subject was elliptic functions. Courant supplemented this material with nine chapters on Riemann surfaces, conformal mapping, and automorphic functions. Courant used an informal, intuitive notion of a surface that displeased some readers but pleased others.

Two years later, in 1924, the first volume of Courant- Hilbert appeared. It was based on lecture notes of Hilbert but even more on Courant's own research in the past five years. The book starts with a 40-page chapter on linear algebra, presented from an analytic point of view, so that generalization to infinite dimension comes naturally. This is followed by chapters on orthogonal function systems, the

Fredholm theory of integral equations, the calculus of variations, and the vibrations of continuum mechanical systems, using extensively the spectral theory of self-adjoint ordinary and partial differential operators. In 1926 Schrödinger invented his wave mechanics, formulated in terms of partial differential operators and their eigenvalues and eigenfunctions. Fortuitously, Courant-Hilbert Volume I contained much of the mathematics needed to understand and solve Schrödinger's equations. This was a striking example of mathematics anticipating the needs of a new physical theory.

In 1928 Courant, Friedrichs, and Lewy published their famous paper on the partial difference equations of mathematical physics. The main motivation for writing it was to use finite difference approximations to prove the existence of solutions of partial differential equations. The paper discusses elliptic, parabolic, and hyperbolic equations; it contains a wealth of ideas, such as the probabilistic interpretation of elliptic difference equations, and the restriction that has to be imposed on the ratio of the time increment and the space increment. The latter, known as the CFL condition, became famous during the computer age. Woe to the computational scientist who ignorantly violates it. This is an outstanding example of research undertaken for purely theoretical purposes turning out to be of immense practical importance.

In 1927 Volume 1 of Courant's calculus text appeared, soon followed by Volume 2. It has been extremely successful in every sense; its translation into English by McShane has sold 50,000 copies of Volume 1 and 35,500 copies of Volume 2 in the United States. It has shaped the minds of many who wanted and needed a deeper grasp of the calculus. Even after 70 years it is better than most, nay all, calculus books in use today in the United States.

In the 1920s and early 1930s Göttingen became again a Mecca of mathematics, as well as of physics. A list of visitors, long term or short term, reads like a Who's Who of mathematicians: Alexandrov, Artin, Birkhoff, Bohr, Hopf, Hardy, Khinchin, Kolmogorov, Lyusternik and Schnirelmann, MacLane, von Neumann, Nielsen, Siegel, Weil, Weyl, Wiener, and many others. Paul Alexandrov described the atmosphere thus: "[Göttingen was] one of the principal centers of world mathematical thought; the place to which all mathematicians came from all over the world, of all possible trends and ages, where there was an exchange of all mathematical ideas and discoveries as soon as they had arisen, no matter where...."

There were many assistants and postdocs around; Courant had private sources of money to pay their stipends. This caused some confusion after the Second World War, when the German government, to its credit, decided to compensate not only faculty members who were dismissed by the Nazis but assistants as well. Many of the assistants in Göttingen thus dismissed had a hard time establishing their claim, for their names did not appear on the roster of those whose salary was paid by the university.

In 1926 concrete negotiations were started, and plans laid, for housing the institutes of mathematics and physics in a permanent building, for long a dream of Felix Klein, now enthusiastically taken up by Courant. The International Educational Board of the Rockefeller Foundation agreed to supply $350,000, and the Prussian Ministry of Education agreed to cover the maintenance costs.

The building of the institute was finished and dedicated in 1929. Courant became its director. Yet this moment of triumph already contained the seed of its own destruction, and that of most civilized Western institutions. The stock market in the United States crashed a few months earlier, leading to a deep economic depression that soon became worldwide. The misery caused by this drove a sizeable part of the German voting population, already embittered by the defeat in the First World War, to support the Nazis. In January 1933 the Hitler gang took over the government. It soon established a new age of barbarism in Germany. For a start, Jewish employees of the state, including professors, were dismissed summarily, Courant among the first. For once his grasp of reality deserted him, and he went from pillar to post to be reinstated. A chance encounter with a member of the Nazi party, who was a member of the university community, set him right. "No doubt you and your friends believe that the excesses of the first few months of the new regime will die down and everything will return to how it was before," said the man. "You are mistaken. Things will get worse and worse for you. You had better get out while you can."

Get out Courant did. After a brief stay in England he, his family, and some family friends landed in New York, where thanks to Oswald Veblen and Abraham Flexner, a position was offered to him in the department of mathematics of New York University, with the charge to develop a graduate program. His host there was the mathematician Donald Flanders, admired by all for his saintly character. He and Courant formed a deep friendship that today extends to their children.

At NYU Courant found a mathematical desert. How he made it bloom is a fascinating story. It started in 1936 with a burst of creative energy. He showed how to solve Plateau's problem-finding a minimal surface spanning a given contour in space-by using Dirichlet's principle. A solution of this classical problem had been found earlier by Jesse Douglas, and in another way by Tibor Radó, but the elegance and simplicity of Courant's method had opened the way to attack more general problems concerning minimal surfaces, such as minimal surfaces spanning multiple contours, of higher genus, and having part of their boundary restricted to a prescribed surface. Courant pursued these generalizations during the next 10 years. The culmination was the book *Dirichlet's Principle, Conformal Mapping, and Minimal Surfaces* that appeared in 1950.

In 1937 Courant was joined at NYU by his brilliant former student from Göttingen, K. O. Friedrichs, and by James J. Stoker, an American. Stoker's original training had been in engineering. In the 1930s, mid-career, he decided to seek a Ph.D. in mechanics at the Federal Institute of Technology in Zürich. One of the first courses he took there was by Heinz Hopf on geometry. Stoker was so charmed by the subject, and the teacher, that he switched his doctoral studies to differential geometry. Hopf wrote to Courant to call his attention to this *junger Amerikaner* whose scientific outlook and temperament were so close to Courant's.

With Friedrich's help Courant was able to complete the long-awaited second volume of Courant-Hilbert. It was, along with Hadamard's book on the Cauchy problem, the first modern text on partial differential equations.

In 1941 *What Is Mathematics?* by Courant and Robbins appeared, a highly popular book written "for beginners and scholars, for students and teachers, for philosophers and engineers, for classrooms and libraries." In the preface Courant warns about the danger facing the traditional place of mathematics in education, and outlines what to do about it. It ought to be compulsory reading for all who today are engaged in reforming the teaching of mathematics.

Courant found in New York City a "vast reservoir" of talented young people, and he was eager to attract them to study mathematics at NYU. To enable those who worked during the day to attend classes, graduate courses were offered in the evening, once a week, for two hours at a time.

America's entry into the Second World War transformed most American academic scientific institutions, none more than Courant's operation at NYU. Government funding was made available for research relevant to war work through the Office of Scientific Research and Development (OSRD). Its head, Vannevar Bush, saw the importance of mathematics for the war effort and set up the Applied Mathematics Panel under the direction of Warren Weaver. Courant was soon invited to be a member of this elite group.

The mathematical project at NYU sponsored by the OSRD was about the flow of compressible fluids in general and the formation and propagation of shock waves in particular. There was enough money to support young research associates (Max Shiffman, Bernard Friedman, and Rudolf Lüneburg), who also served as adjunct faculty in the graduate school. There was also money to provide stipends for graduate students, some of whom were drawn into war work. Courant insisted that graduate training continue even during the war.

This is a good place to describe Courant as a classroom teacher. He seldom bothered to prepare the technical details of his lecture. He muttered in a low voice, and his writing was often indecipherable. Nevertheless, he managed to convey the essence of the subject and left the better students with a warm glow of belief that they could nail down the details better than the master.

Courant supervised many graduate students' doctoral dissertations, more than 20 in Göttingen and a like number at NYU. Among the former were Kurt Friedrichs, Edgar Krahn, Reinhold Baer, Hans Lewy, Otto Neugebauer, Willi Feller, Franz Rellich, Rudolf Lüneburg, Herbert Busemann, and Leifur Asgeirsson. In the United States he taught Max Shiffman, Joe Keller, Harold Grad, Avron Douglis, Martin Kruskal, Anneli Lax, Herbert Kranzer, and Donald Ludwig- a very fine record.

The bulk of the research conducted during the war was fashioned into a book on supersonic flows and shock waves. The editing was in the hands of Cathleen Morawetz, who had the delicate task of reconciling Courant's freely flowing exposition with Friedrichs's demand for precision. *Supersonic Flow and Shock Waves* appeared in book form in 1948; it was a very useful and successful treatise, with a mathematical flavor, on the flow of compressible fluids.

As the war neared its end Courant's thoughts turned to postwar developments. He shrewdly realized that after the war the U.S. government would continue to support science. The critical contribution of science to the war effort had been noted by statesmen: radar, the proximity fuse, bomb sights, code breaking, aerodynamic design, and the atomic bomb. Courant also realized that applied mathematics would be an important part of the government's plans, and he successfully used his wartime contacts to gain support for his vision of applied mathematics. Support came from the office of Naval Research, the Atomic Energy Commission, the offices of Army and Air Force research, and later from the National Science Foundation. As always, he emphasized that research must be combined with teaching.

When it came to hiring faculty, Courant relied on his intuition. The candidate's personality was often more important than the field he was working in. Courant did not like following fashions and fads. "I am against panic buying in an inflated market" was his motto.

Courant did not extend his hatred of Nazis to the German nation. He overcame his resentments and was eager to help those deserving help. He traveled to Germany as soon as it was possible, in 1947, to see the situation there first hand, to talk to people he trusted. He arranged visits to the United States for a number of young mathematicians; this had a tremendous effect psychologically and scientifically and earned Courant the gratitude of the younger generation of German mathematicians and later the Knight Commander's Cross of the Order of Merit of the Federal Republic of Germany.

In 1948 Courant's sixtieth birthday was celebrated with much emotion by mathematicians invited from both sides of the Atlantic; nostalgia flowed like water, held within bounds by Courant's natural irony.

In 1954 the Atomic Energy Commission decided to place one of its supercomputers, the UNIVAC, at a university. After a fierce competition Courant's institute was chosen. The UNIVAC had a memory of 1,000 words and used punched cards. The commission repeatedly replaced it with newer models; the last one was a CDC 6600, installed in 1966 and put out to pasture in 1972.

Courant retired in 1958 at age 70; his successor was Stoker. In retirement Courant succeeded in finishing the translation into English of Volume 2 of *Methods of Mathematical Physics*. This was no mere translation. Courant made a serious attempt, with much help from younger colleagues, to update the material from a mere 470 to over 800 pages. The book ends with a 30-page essay written by Courant on ideal functions, such as distributions. The last chapter in the German original, on existence proofs using variational methods, was omitted. Courant planned to rework it, together with a discussion of finite difference methods, and to issue it as Volume 3. Alas, he was not up to the task.

Courant was deeply concerned about the Cold War. He felt that the natural comradeship of scientists, in particular of mathematicians, might set an example and overcome the "us versus them" stereotypes. Accordingly, he was among the first to visit the Soviet Union. The time—the summer of 1960—was not auspicious, for the Soviets had just shot down a U.S. U-2 spy plane. The remains of the plane and the spy paraphernalia were displayed in the middle of Moscow's Gorky Park. There was a long line of curiosity seekers. As a distinguished visitor, Courant was whisked to the head of the line and was introduced to the aeronautical engineer who was there to explain the workings of the U-2. The engineer was deeply honored: "Professor Courant, I learned aerodynamics from your book." It had been translated in 1950 into Russian, as were all of Courant's other books.

In 1963 Courant led a delegation of about 15 U.S. mathematicians to a two-week conference on the occasion of the opening of the Academic City and University at Novosibirsk. It was a golden time and gave rise to friendships that lasted lifetimes.

A very generous gift from the Sloan Foundation, augmented by the Ford Foundation and the National Science Foundation, was used to construct a handsome 13-story building just off Washington Square, in which the Courant Institute, so named at its dedication in 1965, still nestles. Its architects won all kinds of prizes and went on to build for departments of mathematics at Princeton and Rutgers.

Courant's last years were full of recognition and honors. Solomon Lefschetz admired Courant for having built an enduring school of mathematics, and had nominated him to receive a National Medal of Science. None of these encomiums, however, could lift Courant's spirits in his extreme old age. His institute was thriving, his children and grandchildren were happily launched on careers, but none of that would dissipate his gloom. His old stratagems to overcome depression—embark on a new project, make the acquaintance of a fascinating woman—were no longer available to him. He even stopped playing the piano, which had been a great source of pleasure for him in the past, a way of transcending conflicts and disappointments. He died on January 27, 1972, at the age of 84.

It is time to look back and ask what manner of man was Courant. For this we must look at the testimony of people who knew him intimately. Surprisingly, they were utterly different from Courant in many, sometimes all ways, such as Flanders, a descendent of Puritans and a puritan himself. Flanders was haunted by a lack of confidence. He loved and needed Courant's ebullience, and Flanders's wit and pure spirit were deeply necessary to Courant.

There was Otto Neugebauer, a meticulous and workaholic scholar, about whom Courant said that "he had all the virtues and none of the faults of pedantry." Neugebauer in turn described Courant's style of operation thus:

> All that lies before us as scientific achievement and organization seems to be the outcome of a well conceived plan. To us, nearby, things seemed sometimes more chaotic than planned, and we were far from always in agreement. But a never failing loyalty bound Courant's associates together; he inspired an unshakable confidence in his profound desire to do what was right and what made sense under the given circumstances. His ability to create a feeling of mutual confidence in those who know him intimately lies at the foundation of his success and influence.

Friedrichs was another former student utterly devoted to Courant, although totally different from him. He described the excitement he felt when as a young student he read Courant's presentation of geometric function theory in Hurwitz-Courant.

> It is true that there were some passages in which matters of rigor were taken somewhat lightly, but the essence came through marvelously. I was reminded of this effect much later, when I heard Courant play some Beethoven piano sonata. There were also some difficult passages which he somehow simplified; but the essence carried over wonderfully. In a way, one could perhaps say the same thing about his skiing—a sport which, incidentally, his assistants were expected to have mastered, or else to learn from him. Never mind the details of the operations, he always managed to come down the mountain quite safely.

Here are the observations of another close friend, Lucile Gardner Wolff.

> There are some great men—and among them some of the greatest—who owe their preeminence not merely to their good qualities but to their bad, or to what would have been bad in another man; whose talents, however remarkable, cannot in themselves account for their achievements or explain why they succeeded where others, equally gifted, failed; whose genius lay precisely

in the ability to turn their weak points to good account. Such a one is the late Richard Courant.

Again Friedrichs:

> As a person Richard Courant cannot be measured by any common standard. Think of it: a mathematician who hated logic, who abhorred abstractions, who was suspicious of truth—if it was just bare truth. For a mathematician these seem to be contradictions. But Richard Courant was never afraid of contradictions—if they could enhance the fullness of life.

Courant loved to be with young people. He understood their ambitions and anxieties and was ready with useful, often unconventional advice. For many the Courant Institute was a second family, and some of this spirit abides today.

Courant has been gone for nearly 30 years, surely enough time for the verdict of history. What is remembered? His insistence on the fundamental unity of all mathematical disciplines and on the vital connection between mathematics and other sciences. The name he gave to his institute—Institute of Mathematical Sciences—expressed his attitude. It is remarkable that today many of the leading mathematical institutions have adopted this appellation.

Courant insisted that research be combined with teaching, a philosophy he liked to trace back to the French Revolution and the founding of the *Ecole Polytechnique.*

Courant was a superb writer, in both German and English. Three of his books—*What Is Mathematics?*, his two volumes of calculus, and *Supersonic Flow and Shock Waves*, written with Friedrichs-are alive and well. Even his occasional pieces are worth re-reading. For example, in an article on mathematics in the modern world, which appeared in 1964 in *Scientific American*, he wrote,

> To handle the translation of reality into the abstract models of mathematics and to appraise the degree of accuracy thereby attainable calls for intuitive feelings sharpened by experience. It may also often involve the framing of genuine mathematical problems that are far too difficult to be solved by the available techniques of the science. Such, in part, is the nature of the intellectual adventure and the satisfaction experienced by the mathematician who works with engineers and natural scientists on the mastering of the real problems that arise in so many places as man extends his understanding and control of nature.

Those who wish to find out more about Courant's adventurous life can learn much from Constance Reid's biography subtitled "The Story of an Improbable Mathematician." Those who knew him remember his great warmth and kindness, often disguised by irony, his energy and enthusiasm coupled with skepticism, and his inexhaustible optimism in the face of seemingly insurmountable obstacles.

Selected Bibliography

Beweis des Satzes, dass von allen homogenen Membranen gegebenen Umfantes und gegebener Spannung die kreisförmige den tiefsten Grundton besizt. *Math. Z.* 3:321-28, 1918.

Über die Eigenwerte bei den Differenzialgleichungen der mathematischen Physik. *Math. Z.* 7:1-57, 1920.

Hurwitz-Courant, Vorlesunger über allgemeine Funcktionen Theorie (4th ed., with an appendix by H. Röhrl, vol. 3, Grundlehren der mathematischen Wissenschaften, Springer, 1964), originally published in 1922.

With D. Hilbert. *Methoden der Mathematischen Physik. Vol. 1*, Springer Verlag, 1924 (Vol. 2, Springer, 1937.)

With K. O. Friedrichs and H. Lewy. Über die partiellen Differenzengleichungen der Mathematischen Physik. *Math. Ann.* 100:32-74, 1928.

Differential and Integral Calculus. Vol. 1, Translated by E. J. McShane, Nordemann, 1934; *Vol. 2*, Nordemann, 1936.

Plateau's problem and Dirichlet's principle. *Ann. Math.* 38(2):679 724, 1937.

The existence of minimal surfaces of given topological structure under given boundary conditions. *Acta Math.* 72:51-98, 1940.

1941 With H. Robbins. *What Is Mathematics?* An elementary approach to ideas and methods, Oxford University Press, 1941 (2nd ed., revised by Ian Stewart. Oxford University Press, 1996.)

With K. O. Friedrichs. *Supersonic Flow and Shock Waves*. Wiley-Interscience, 1948.

APPENDIX 4

Peter D. Lax curriculum vitae

Peter D. Lax
New York University
Courant Institute of Mathematical Sciences
251 Mercer Street
New York, NY 10012
Born: May 1, 1926
Budapest, Hungary

EDUCATION:
New York University, A.B., 1947
New York University, Ph.D., 1949

POSITIONS:
Los Alamos Scientific Laboratory, 1945-46
Manhattan Project
Staff Member, 1950, Los Alamos Scientific Laboratory
Assistant Professor, 1951, New York University
Fulbright Lecturer in Germany, 1958
Professor, 1958–Present, New York University
Director, 1972–80, Courant Institute of Mathematical Sciences, New York University

HONORS AND AWARDS:
Lester R. Ford, 1966, 1973
von Neumann Lecturer, SIAM, 1969
Hermann Weyl Lecturer, 1972
Hedrick Lecturer, 1973
Chauvenet Prize, Mathematical Association of America, 1974
Norbert Wiener Prize, American Mathematical
Society and Society of Industrial and Applied Mathematics, 1975
Member, National Academy of Sciences of the U.S.A.
Member, American Academy of Arts and Sciences
Honorary Life Member, New York Academy of Sciences, 1982
Foreign Associate, French Academy of Sciences
National Academy of Sciences
Award in Applied Mathematics and Numerical Sciences, 1983
National Medal of Science, 1986
Wolf Prize, 1987
Member, Soviet Academy of Sciences, 1989
Steele Prize, 1992

Member, Hungarian Academy of Sciences, 1993
Member, Academia Sinica, Beijing, 1993
Distinguished Teaching Award, New York University, 1995
Member, Moscow Mathematical Society, 1995
Abel Prize, 2005

HONORARY DOCTORAL DEGREES:
Kent State University, 1975
University of Paris, 1979
Technical University of Aachen, 1988
Heriot-Watt University, 1990
Tel Aviv University, 1992
University of Maryland, Baltimore, 1993
Brown University, 1993
Beijing University, 1993
Texas A&M University, 2000
University of Pennsylvania, 2012
Tulane University, 2012

PROFESSIONAL SOCIETIES:
Board of Governors, Mathematical Association of America, 1966–67
New York Academy of Sciences, 1986–87
Member, Society of Industrial and Applied Mathematics
Vice President, American Mathematical Society, 1969–71
President, American Mathematical Society, 1977–80

GOVERNMENT SERVICE:
President's Committee on the National Medal of Science, 1977
National Science Board, 1980–86
DOE Related:
Theory Division, Advisory Committee, LANL
Senior Fellow, Los Alamos Scientific Laboratory
Review Committee, Oak Ridge National Laboratory

List of Publications

Peter D. Lax, *Numbers and functions: from a classical experimental mathematicians point of view [book review of MR2963308]*, Amer. Math. Monthly **121** (2014), no. 2, 183. MR3202890

Jerry L. Kazdan, Peter D. Lax, Albert B. J. Novikoff, C. Denson Hill, Antony Jameson, Eva V. Swenson, Ruth Bers Shapiro, Cathy Garabedian, and Emily Garabedian, *Paul Roesel Garabedian (1927–2010)*, Notices Amer. Math. Soc. **61** (2014), no. 3, 244–255, DOI 10.1090/noti1088. MR3185358

Peter D. Lax and Maria Shea Terrell, *Calculus with applications*, 2nd ed., Undergraduate Texts in Mathematics, Springer, New York, 2014. MR3113606

Peter D. Lax, *Selected papers. II*, Springer Collected Works in Mathematics, Springer, New York, 2013. Edited by Peter Sarnak and Andrew J. Majda; Reprint of the 2005 edition [MR2164868]. MR3185091

———, *Selected papers. I*, Springer Collected Works in Mathematics, Springer, New York, 2013. Edited by Peter Sarnak and Andrew J. Majda; Reprint of the 2005 edition [MR2164867]. MR3185090

———, *Stability of difference schemes*, The Courant-Friedrichs-Lewy (CFL) condition, Birkhäuser/Springer, New York, 2013, pp. 1–7, DOI 10.1007/978-0-8176-8394-8_1. MR3050166

Peter D. Lax and Lawrence Zalcman, *Complex proofs of real theorems*, University Lecture Series, vol. 58, American Mathematical Society, Providence, RI, 2012. MR2827550

Elisha Falbel, Gábor Francsics, Peter D. Lax, and John R. Parker, *Generators of a Picard modular group in two complex dimensions*, Proc. Amer. Math. Soc. **139** (2011), no. 7, 2439–2447, DOI 10.1090/S0002-9939-2010-10653-6. MR2784810 (2012d:22014)

Peter D. Lax, Friedrich Hirzebruch, Barry Mazur, Lawrence Conlon, Edward B. Curtis, Harold M. Edwards, Johannes Huebschmann, and Herbert Shulman, *Raoul Bott as we knew him*, A celebration of the mathematical legacy of Raoul Bott, CRM Proc. Lecture Notes, vol. 50, Amer. Math. Soc., Providence, RI, 2010, pp. 43–49. MR2648884

Peter D. Lax, *Rethinking the Lebesgue integral*, Amer. Math. Monthly **116** (2009), no. 10, 863–881, DOI 10.4169/000298909X476998. MR2589217 (2010k:26010)

———, *A curious functional equation*, J. Anal. Math. **105** (2008), 383–390, DOI 10.1007/s11854-008-0042-4. MR2438432 (2010b:39032)

———, *Mathematics and physics*, Bull. Amer. Math. Soc. (N.S.) **45** (2008), no. 1, 135–152 (electronic), DOI 10.1090/S0273-0979-07-01182-2. MR2358380

———, *Linear algebra and its applications*, 2nd ed., Pure and Applied Mathematics (Hoboken), Wiley-Interscience [John Wiley & Sons], Hoboken, NJ, 2007. MR2356919 (2008j:15002)

———, *The Cauchy integral theorem*, Amer. Math. Monthly **114** (2007), no. 8, 725–727. MR2354441

———, *Computational fluid dynamics*, J. Sci. Comput. **31** (2007), no. 1-2, 185–193, DOI 10.1007/s10915-006-9104-x. MR2304275

Peter Lax, *Thoughts on Gaetano Fichera*, Rend. Accad. Naz. Sci. XL Mem. Mat. Appl. (5) **30** (2006), 1. MR2489588

Peter D. Lax, *Gibbs phenomena*, J. Sci. Comput. **28** (2006), no. 2-3, 445–449, DOI 10.1007/s10915-006-9075-y. MR2272639 (2007j:65082)

———, *Hyperbolic partial differential equations*, Courant Lecture Notes in Mathematics, vol. 14, New York University, Courant Institute of Mathematical Sciences, New York; American Mathematical Society, Providence, RI, 2006. With an appendix by Cathleen S. Morawetz. MR2273657 (2007h:35002)

Gábor Francsics and Peter D. Lax, *Analysis of a Picard modular group*, Proc. Natl. Acad. Sci. USA **103** (2006), no. 30, 11103–11105 (electronic), DOI 10.1073/pnas.0603075103. MR2242649 (2007h:11048)

Peter D. Lax, *Selected papers. Vol. II*, Springer, New York, 2005. Edited by Peter Sarnak and Andrew Majda. MR2164868 (2006k:35001b)

———, *Selected papers. Vol. I*, Springer, New York, 2005. Edited by Peter Sarnak and Andrew Majda. MR2164867 (2006k:35001a)

Gábor Francsics and Peter D. Lax, *A semi-explicit fundamental domain for a Picard modular group in complex hyperbolic space*, Geometric analysis of PDE and several complex variables, Contemp. Math., vol. 368, Amer. Math. Soc., Providence, RI, 2005, pp. 211–226, DOI 10.1090/conm/368/06780. MR2126471 (2006b:22011)

Susan Friedlander, Peter Lax, Cathleen Morawetz, Louis Nirenberg, Gregory Seregin, Nina Ural′tseva, and Mark Vishik, *Olga Alexandrovna Ladyzhenskaya (1922–2004)*, Notices Amer. Math. Soc. **51** (2004), no. 11, 1320–1331. MR2105237

V. Eger, P. Laks, and K. Moravets, *Ol′ga Arsen′evna Oleĭnik* (Russian), Tr. Semin. im. I. G. Petrovskogo **23** (2003), 7–15, DOI 10.1023/B:JOTH.0000016046.59508.ff; English transl., J. Math. Sci. (N. Y.) **120** (2004), no. 3, 1242–1246. MR2085178

Peter D. Lax, *From the early days of the theory of distributions*, Gaz. Math. **98, suppl.** (2003), 63–64. Laurent Schwartz (1915–2002). MR2067350

———, *Max Shiffman (1914–2000)*, Notices Amer. Math. Soc. **50** (2003), no. 11, 1401. MR2011607

Xu-Dong Liu and Peter D. Lax, *Positive schemes for solving multi-dimensional hyperbolic systems of conservation laws. II*, J. Comput. Phys. **187** (2003), no. 2, 428–440, DOI 10.1016/S0021-9991(03)00100-1. MR1980267 (2004e:65097)

Willi Jäger, Peter Lax, and Cathleen Synge Morawetz, *Olga Arsen′evna Oleĭnik (1925–2001)*, Notices Amer. Math. Soc. **50** (2003), no. 2, 220–223. MR1951108 (2003k:01029)

Peter D. Lax, *Jürgen Moser, 1928–1999*, Ergodic Theory Dynam. Systems **22** (2002), no. 5, 1337–1342, DOI 10.1017/S0143385702001050. MR1934139

———, *Functional analysis*, Pure and Applied Mathematics (New York), Wiley-Interscience [John Wiley & Sons], New York, 2002. MR1892228 (2003a:47001)

Peter D. Lax, Enrico Magenes, and Roger Temam, *Jacques-Louis Lions (1928–2001)*, Notices Amer. Math. Soc. **48** (2001), no. 11, 1315–1321. MR1870634

Peter D. Lax, *The Radon transform and translation representation*, J. Evol. Equ. **1** (2001), no. 3, 311–323, DOI 10.1007/PL00001373. Dedicated to Ralph S. Phillips. MR1861225 (2003a:58047)

Lennart Carleson, Sun-Yung Alice Chang, Peter W. Jones, Markus Keel, Peter D. Lax, Nikolai Makarov, Donald Sarason, Wilhelm Schlag, and Barry Simon, *Thomas H. Wolff (1954–2000)*, Notices Amer. Math. Soc. **48** (2001), no. 5, 482–490. MR1822959 (2001m:01041)

Peter D. Lax, *Change of variables in multiple integrals. II*, Amer. Math. Monthly **108** (2001), no. 2, 115–119, DOI 10.2307/2695524. MR1818184 (2001m:26028)

———, *On the accuracy of Glimm's scheme*, Methods Appl. Anal. **7** (2000), no. 3, 473–477. Cathleen Morawetz: a great mathematician. MR1869298 (2002k:65166)

———, *Mathematics and computing*, Mathematics: frontiers and perspectives, Amer. Math. Soc., Providence, RI, 2000, pp. 417–432. MR1754789 (2001e:65001)

Armand Borel, Gennadi M. Henkin, and Peter D. Lax, *Jean Leray (1906–1998)*, Notices Amer. Math. Soc. **47** (2000), no. 3, 350–359. MR1740391 (2000j:01048)

Stefan Hildebrandt and Peter D. Lax, *Otto Toeplitz* (German), Bonner Mathematische Schriften [Bonn Mathematical Publications], 319, Universität Bonn, Mathematisches Institut, Bonn, 1999. MR1944435 (2004d:01033)

Peter D. Lax, *Change of variables in multiple integrals*, Amer. Math. Monthly **106** (1999), no. 6, 497–501, DOI 10.2307/2589462. MR1699248 (2000c:26010)

———, *The beginnings of applied mathematics after the Second World War*, Quart. Appl. Math. **56** (1998), no. 4, 607–615. Current and future challenges in the applications of mathematics (Providence, RI, 1997). MR1668731

Ami Harten, Peter D. Lax, C. David Levermore, and William J. Morokoff, *Convex entropies and hyperbolicity for general Euler equations*, SIAM J. Numer. Anal. **35** (1998), no. 6, 2117–2127 (electronic), DOI 10.1137/S0036142997316700. MR1655839 (99j:76107)

Peter D. Lax, *On the discriminant of real symmetric matrices*, Comm. Pure Appl. Math. **51** (1998), no. 11-12, 1387–1396, DOI 10.1002/(SICI)1097-0312(199811/12)51:11/12¡1387::AID-CPA6¿3.3.CO;2-S. MR1639147 (99i:15010)

Peter D. Lax and Xu-Dong Liu, *Solution of two-dimensional Riemann problems of gas dynamics by positive schemes*, SIAM J. Sci. Comput. **19** (1998), no. 2, 319–340 (electronic), DOI 10.1137/S1064827595291819. MR1618863

Peter D. Lax, *The zero dispersion limit, a deterministic analogue of turbulence*, Nonlinear evolutionary partial differential equations (Beijing, 1993), AMS/IP Stud. Adv. Math., vol. 3, Amer. Math. Soc., Providence, RI, 1997, pp. 53–64. MR1468482 (98i:35166)

―――, *Linear algebra*, Pure and Applied Mathematics (New York), John Wiley & Sons, Inc., New York, 1997. A Wiley-Interscience Publication. MR1423602 (99c:15001)

―――, *Outline of a theory of the KdV equation*, Recent mathematical methods in nonlinear wave propagation (Montecatini Terme, 1994), Lecture Notes in Math., vol. 1640, Springer, Berlin, 1996, pp. 70–102, DOI 10.1007/BFb0093707. MR1600908 (98m:35184)

G. Bolliat, C. M. Dafermos, P. D. Lax, and T. P. Liu, *Recent mathematical methods in nonlinear wave propagation*, Lecture Notes in Mathematics, vol. 1640, Springer-Verlag, Berlin; Centro Internazionale Matematico Estivo (C.I.M.E.), Florence, 1996. Lectures given at the 1st C.I.M.E. Session held in Montecatini Terme, May 23–31, 1994; Edited by T. Ruggeri; Fondazione C.I.M.E.. [C.I.M.E. Foundation]. MR1600896 (98g:35002)

Xu-Dong Liu and Peter D. Lax, *Positive schemes for solving multi-dimensional hyperbolic systems of conservation laws*, Proceedings of the VIII International Conference on Waves and Stability in Continuous Media, Part I (Palermo, 1995), Rend. Circ. Mat. Palermo (2) Suppl. **45** (1996), 367–375. MR1461086 (98e:65065)

Peter Lax, *Erratum: "Ami Harten [?–1994]" [Comm. Pure Appl. Math. 48 (1995), no. 12, 1303]*, Comm. Pure Appl. Math. **49** (1996), no. 8, 867. MR1391758

―――, *Ami Harten [?–1994]*, Comm. Pure Appl. Math. **48** (1995), no. 12, 1303. MR1369390

Peter D. Lax, *A short path to the shortest path*, Amer. Math. Monthly **102** (1995), no. 2, 158–159, DOI 10.2307/2975350. MR1315595

―――, *Cornelius Lanczos (1893–1974), and the Hungarian phenomenon in science and mathematics*, Proceedings of the Cornelius Lanczos International Centenary Conference (Raleigh, NC, 1993), SIAM, Philadelphia, PA, 1994, pp. xlix–lii. MR1298214

―――, *Trace formulas for the Schroedinger operator*, Comm. Pure Appl. Math. **47** (1994), no. 4, 503–512, DOI 10.1002/cpa.3160470405. MR1272386 (95c:34144)

P. D. Lax, C. D. Levermore, and S. Venakides, *The generation and propagation of oscillations in dispersive initial value problems and their limiting behavior*, Important developments in soliton theory, Springer Ser. Nonlinear Dynam., Springer, Berlin, 1993, pp. 205–241. MR1280476 (95c:35245)

Peter D. Lax, *The existence of eigenvalues of integral operators*, Indiana Univ. Math. J. **42** (1993), no. 3, 889–891, DOI 10.1512/iumj.1993.42.42040. MR1254123 (94i:47080)

Peter D. Lax and Ralph S. Phillips, *Translation representation for automorphic solutions of the wave equation in non-Euclidean spaces. IV*, Comm. Pure Appl. Math. **45** (1992), no. 2, 179–201, DOI 10.1002/cpa.3160450203. MR1139065 (93c:58220)

Peter D. Lax, *The zero dispersion limit, a deterministic analogue of turbulence*, Comm. Pure Appl. Math. **44** (1991), no. 8-9, 1047–1056, DOI 10.1002/cpa.3160440815. MR1127048 (92j:35151)

P. D. Lax, *Deterministic theories of turbulence*, Frontiers in pure and applied mathematics, North-Holland, Amsterdam, 1991, pp. 179–184. MR1110599 (92i:76058)

Thomas Y. Hou and Peter D. Lax, *Dispersive approximations in fluid dynamics*, Comm. Pure Appl. Math. **44** (1991), no. 1, 1–40, DOI 10.1002/cpa.3160440102. MR1077912 (91m:76088)

Peter D. Lax, *Book Review: From cardinals to chaos: Reflections on the life and legacy of Stan Ulam*, Bull. Amer. Math. Soc. (N.S.) **22** (1990), no. 2, 304–310, DOI 10.1090/S0273-0979-1990-15893-8. MR1567846

―――, *Remembering John von Neumann*, The legacy of John von Neumann (Hempstead, NY, 1988), Proc. Sympos. Pure Math., vol. 50, Amer. Math. Soc., Providence, RI, 1990, pp. 5–7, DOI 10.1090/pspum/050/1067745. MR1067745 (92e:01066)

―――, *The ergodic character of sequences of pedal triangles*, Amer. Math. Monthly **97** (1990), no. 5, 377–381, DOI 10.2307/2324387. MR1048909 (91h:51028)

_____, *The flowering of applied mathematics in America*, AMS-MAA Joint Lecture Series, American Mathematical Society, Providence, RI, 1989. A joint AMS-MAA invited address presented in Providence, Rhode Island, August 1988. MR1056291 (92c:01035)

Peter D. Lax and Ralph S. Phillips, *Scattering theory*, 2nd ed., Pure and Applied Mathematics, vol. 26, Academic Press, Inc., Boston, MA, 1989. With appendices by Cathleen S. Morawetz and Georg Schmidt. MR1037774 (90k:35005)

Peter D. Lax, *The flowering of applied mathematics in America*, SIAM Rev. **31** (1989), no. 4, 533–541, DOI 10.1137/1031123. Reprinted from *A century of mathematics, Part II*, 455–466, Amer. Math. Soc., Providence, RI, 1989 [see MR1003117 (90a:01065)]. MR1025479

_____, *The flowering of applied mathematics in America*, A century of mathematics in America, Part II, Hist. Math., vol. 2, Amer. Math. Soc., Providence, RI, 1989, pp. 455–466. MR1003149 (90m:01037)

_____, *Mathematics and computing*, ICIAM '87: Proceedings of the First International Conference on Industrial and Applied Mathematics (Paris, 1987), SIAM, Philadelphia, PA, 1988, pp. 137–143. MR976856

Jonathan Goodman and Peter D. Lax, *On dispersive difference schemes. I*, Comm. Pure Appl. Math. **41** (1988), no. 5, 591–613, DOI 10.1002/cpa.3160410506. MR948073 (89f:65094)

Peter D. Lax, *Oscillatory solutions of partial differential and difference equations*, Mathematics applied to science (New Orleans, La., 1986), Academic Press, Boston, MA, 1988, pp. 155–170. MR934947 (89i:35006)

Randall J. LeVeque, Charles S. Peskin, and Peter D. Lax, *Solution of a two-dimensional cochlea model with fluid viscosity*, SIAM J. Appl. Math. **48** (1988), no. 1, 191–213, DOI 10.1137/0148009. MR923297 (89b:92007)

Peter D. Lax, *Mathematics and its applications* (Czech), Pokroky Mat. Fyz. Astronom. **32** (1987), no. 6, 309–314. Translated from the English by O. Kowalski. MR926883 (89b:00038)

_____, *On symmetrizing hyperbolic differential equations*, Nonlinear hyperbolic problems (St. Etienne, 1986), Lecture Notes in Math., vol. 1270, Springer, Berlin, 1987, pp. 150–151, DOI 10.1007/BFb0078324. MR910111

A. Jameson and P. D. Lax, *Corrigendum: "Conditions for the construction of multipoint total variation diminishing difference schemes"*, Appl. Numer. Math. **3** (1987), no. 3, 289, DOI 10.1016/0168-9274(87)90053-5. MR898060 (88f:65138b)

Peter D. Lax, *Hyperbolic systems of conservation laws in several space variables*, Current topics in partial differential equations, Kinokuniya, Tokyo, 1986, pp. 327–341. MR1112153

_____, *On the weak convergence of dispersive difference schemes*, Oscillation theory, computation, and methods of compensated compactness (Minneapolis, Minn., 1985), IMA Vol. Math. Appl., vol. 2, Springer, New York, 1986, pp. 107–113, DOI 10.1007/978-1-4613-8689-6_5. MR869823

Antony Jameson and Peter D. Lax, *Conditions for the construction of multipoint total variation diminishing difference schemes*, Appl. Numer. Math. **2** (1986), no. 3-5, 335–345, DOI 10.1016/0168-9274(86)90038-3. MR863992 (88f:65138a)

Peter D. Lax, *Mathematics and its applications*, Math. Intelligencer **8** (1986), no. 4, 14–17, DOI 10.1007/BF03026113. MR858296 (88e:00017)

_____, *Mathematics and computing*, Proceedings of the conference on frontiers of quantum Monte Carlo (Los Alamos, N.M., 1985), J. Statist. Phys. **43** (1986), no. 5-6, 749–756, DOI 10.1007/BF02628302. MR854281 (87j:00038)

_____, *On dispersive difference schemes*, Phys. D **18** (1986), no. 1-3, 250–254, DOI 10.1016/0167-2789(86)90185-5. Solitons and coherent structures (Santa Barbara, Calif., 1985). MR838330 (87h:65155)

P. Constantin, P. D. Lax, and A. Majda, *A simple one-dimensional model for the three-dimensional vorticity equation*, Comm. Pure Appl. Math. **38** (1985), no. 6, 715–724, DOI 10.1002/cpa.3160380605. MR812343 (87a:76037)

Randall J. LeVeque, Charles S. Peskin, and Peter D. Lax, *Solution of a two-dimensional cochlea model using transform techniques*, SIAM J. Appl. Math. **45** (1985), no. 3, 450–464, DOI 10.1137/0145026. MR787564 (86i:92006)

Peter D. Lax and Ralph S. Phillips, *Translation representations for automorphic solutions of the wave equation in non-Euclidean spaces; the case of finite volume*, Trans. Amer. Math. Soc. **289** (1985), no. 2, 715–735, DOI 10.2307/2000260. MR784011 (86f:11045)

———, *Translation representations for automorphic solutions of the wave equation in non-Euclidean spaces. III*, Comm. Pure Appl. Math. **38** (1985), no. 2, 179–207, DOI 10.1002/cpa.3160380205. MR780072 (86j:58150)

Peter D. Lax, *Problems solved and unsolved concerning linear and nonlinear partial differential equations*, Proceedings of the International Congress of Mathematicians, Vol. 1, 2 (Warsaw, 1983), PWN, Warsaw, 1984, pp. 119–137. MR804680 (87k:35003)

———, *The zero dispersion limit for the KdV equation*, Differential equations (Birmingham, Ala., 1983), North-Holland Math. Stud., vol. 92, North-Holland, Amsterdam, 1984, pp. 387–390, DOI 10.1016/S0304-0208(08)73719-8. MR799374 (87c:35145)

———, *Shock waves, increase of entropy and loss of information*, Seminar on nonlinear partial differential equations (Berkeley, Calif., 1983), Math. Sci. Res. Inst. Publ., vol. 2, Springer, New York, 1984, pp. 129–171, DOI 10.2172/5648733. MR765233 (86b:35130)

Peter D. Lax and Ralph S. Phillips, *Translation representations for automorphic solutions of the wave equation in non-Euclidean spaces. II*, Comm. Pure Appl. Math. **37** (1984), no. 6, 779–813, DOI 10.1002/cpa.3160370604. MR762873 (86h:58140)

———, *Translation representation for automorphic solutions of the wave equation in non-Euclidean spaces. I*, Comm. Pure Appl. Math. **37** (1984), no. 3, 303–328, DOI 10.1002/cpa.3160370304. MR739923 (86c:58148)

Peter D. Lax and C. David Levermore, *The small dispersion limit of the Korteweg-de Vries equation. III*, Comm. Pure Appl. Math. **36** (1983), no. 6, 809–829, DOI 10.1002/cpa.3160360606. MR720595 (85g:35105c)

———, *The small dispersion limit of the Korteweg-de Vries equation. II*, Comm. Pure Appl. Math. **36** (1983), no. 5, 571–593, DOI 10.1002/cpa.3160360503. MR716197 (85g:35105b)

———, *The small dispersion limit of the Korteweg-de Vries equation. I*, Comm. Pure Appl. Math. **36** (1983), no. 3, 253–290, DOI 10.1002/cpa.3160360302. MR697466 (85g:35105a)

Amiram Harten, Peter D. Lax, and Bram van Leer, *On upstream differencing and Godunov-type schemes for hyperbolic conservation laws*, SIAM Rev. **25** (1983), no. 1, 35–61, DOI 10.1137/1025002. MR693713 (85h:65188)

Peter D. Lax and C. David Levermore, *The zero dispersion limit for the Korteweg-de Vries (KdV) equation*, Proceedings of the 1980 Beijing Symposium on Differential Geometry and Differential Equations, Vol. 1, 2, 3 (Beijing, 1980), Science Press, Beijing, 1982, pp. 717–729. MR714345 (85e:35112)

Peter D. Lax and Ralph S. Phillips, *The asymptotic distribution of lattice points in Euclidean and non-Euclidean spaces*, Toeplitz centennial (Tel Aviv, 1981), Operator Theory: Adv. Appl., vol. 4, Birkhäuser, Basel-Boston, Mass., 1982, pp. 365–375. MR669919 (83m:10089)

———, *The asymptotic distribution of lattice points in Euclidean and non-Euclidean spaces*, J. Funct. Anal. **46** (1982), no. 3, 280–350, DOI 10.1016/0022-1236(82)90050-7. MR661875 (83j:10057)

———, *A local Paley-Wiener theorem for the Radon transform of L_2 functions in a non-Euclidean setting*, Comm. Pure Appl. Math. **35** (1982), no. 4, 531–554, DOI 10.1002/cpa.3160350404. MR657826 (83i:43016)

Peter D. Lax, *The multiplicity of eigenvalues*, Bull. Amer. Math. Soc. (N.S.) **6** (1982), no. 2, 213–214, DOI 10.1090/S0273-0979-1982-14983-7. MR640948 (83a:15009)

Peter D. Lax and Ralph S. Phillips, *The asymptotic distribution of lattice points in Euclidean and non-Euclidean spaces*, Functional analysis and approximation (Oberwolfach, 1980), Internat. Ser. Numer. Math., vol. 60, Birkhäuser, Basel-Boston, Mass., 1981, pp. 373–383. MR650290 (83m:10088)

———, *A local Paley-Wiener theorem for the Radon transform in real hyperbolic spaces*, Mathematical analysis and applications, Part B, Adv. in Math. Suppl. Stud., vol. 7, Academic Press, New York-London, 1981, pp. 483–487. MR634254 (83i:43015)

Peter D. Lax, *Applied mathematics, 1945 to 1975*, American mathematical heritage: algebra and applied mathematics (El Paso, Tex., 1975/Arlington, Tex., 1976), Math. Ser., vol. 13, Texas Tech Univ., Lubbock, Tex., 1981, pp. 95–100. MR641704 (83d:01047)

Amiram Harten and Peter D. Lax, *A random choice finite difference scheme for hyperbolic conservation laws*, SIAM J. Numer. Anal. **18** (1981), no. 2, 289–315, DOI 10.1137/0718021. MR612144 (83b:65090)

Peter D. Lax and Ralph S. Phillips, *The translation representation theorem*, Integral Equations Operator Theory **4** (1981), no. 3, 416–421, DOI 10.1007/BF01697973. MR623545 (82h:47041)

———, *Translation representations for the solution of the non-Euclidean wave equation. II*, Comm. Pure Appl. Math. **34** (1981), no. 3, 347–358, DOI 10.1002/cpa.3160340304. MR611749 (82f:43010)

P. D. Lax and R. S. Phillips, *Correction to: "Translation representations for the solution of the non-Euclidean wave equation" [Comm. Pure Appl. Math. **32** (1979), no. 5, 617–667; MR 81a:43013]*, Comm. Pure Appl. Math. **33** (1980), no. 5, 685, DOI 10.1002/cpa.3160330508. MR586418 (82a:43010)

Peter D. Lax, *On the notion of hyperbolicity*, Comm. Pure Appl. Math. **33** (1980), no. 3, 395–397, DOI 10.1002/cpa.3160330309. MR562741 (81f:35077)

Peter D. Lax and Ralph S. Phillips, *Scattering theory for automorphic functions*, Bull. Amer. Math. Soc. (N.S.) **2** (1980), no. 2, 261–295, DOI 10.1090/S0273-0979-1980-14735-7. MR555264 (81c:10037)

———, *Translation representations for the solution of the non-Euclidean wave equation*, Comm. Pure Appl. Math. **32** (1979), no. 5, 617–667, DOI 10.1002/cpa.3160320503. MR533296 (81a:43013)

Peter D. Lax, *Recent methods for computing discontinuous solutions—a review*, Computing methods in applied sciences and engineering (Proc. Third Internat. Sympos., Versailles, 1977), Lecture Notes in Phys., vol. 91, Springer, Berlin-New York, 1979, pp. 3–12. MR540126 (80j:65039)

Peter D. Lax and C. David Levermore, *The zero dispersion limit for the Korteweg-de Vries KdV equation*, Proc. Nat. Acad. Sci. U.S.A. **76** (1979), no. 8, 3602–3606. MR540258 (80g:35113)

Peter D. Lax and Ralph S. Phillips, *The time delay operator and a related trace formula*, Topics in functional analysis (essays dedicated to M. G. Kreĭn on the occasion of his 70th birthday), Adv. in Math. Suppl. Stud., vol. 3, Academic Press, New York-London, 1978, pp. 197–215. MR538021 (80j:47010)

Peter D. Lax, *The numerical solution of the equations of fluid dynamics*, Lectures on combustion theory (Courant Inst., New York Univ., New York, 1977), New York Univ., New York, 1978, pp. 1–60. MR522092 (80i:76028)

———, *A Hamiltonian approach to the KdV and other equations*, Nonlinear evolution equations (Proc. Sympos., Univ. Wisconsin, Madison, Wis., 1977), Publ. Math. Res. Center Univ. Wisconsin, vol. 40, Academic Press, New York-London, 1978, pp. 207–224. MR513820 (80d:35129)

———, *Chemical kinetics*, Lectures on combustion theory (Courant Inst., New York Univ., New York, 1977), New York Univ., New York, 1978, pp. 122–136. MR522094 (80c:80010)

———, *Accuracy and resolution in the computation of solutions of linear and nonlinear equations*, Recent advances in numerical analysis (Proc. Sympos., Math. Res. Center, Univ. Wisconsin, Madison, Wis., 1978), Publ. Math. Res. Center Univ. Wisconsin, vol. 41, Academic Press, New York-London, 1978, pp. 107–117. MR519059 (80b:65147)

Peter D. Lax and Ralph S. Phillips, *Scattering theory for domains with non-smooth boundaries*, Arch. Rational Mech. Anal. **68** (1978), no. 2, 93–98, DOI 10.1007/BF00281403. MR505506 (80a:35098)

Michael S. Mock and Peter D. Lax, *The computation of discontinuous solutions of linear hyperbolic equations*, Comm. Pure Appl. Math. **31** (1978), no. 4, 423–430. MR0468216 (57 #8054)

Peter D. Lax and Ralph S. Phillips, *An example of Huygens' principle*, Comm. Pure Appl. Math. **31** (1978), no. 4, 415–421. MR0466977 (57 #6850)

Anneli Lax and Peter D. Lax, *On sums of squares*, Linear Algebra and Appl. **20** (1978), no. 1, 71–75. MR0463112 (57 #3074)

Peter D. Lax, *A Hamiltonian approach to the KdV and other equations*, Group theoretical methods in physics (Proc. Fifth Internat. Colloq., Univ. Montréal, Montreal, Que., 1976), Academic Press, New York, 1977, pp. 39–57. MR650117 (58 #31222)

Peter D. Lax and Ralph S. Phillips, *The scattering of sound waves by an obstacle*, Comm. Pure Appl. Math. **30** (1977), no. 2, 195–233. MR0442510 (56 #892)

Peter D. Lax, *Almost periodic behavior of nonlinear waves*, Surveys in applied mathematics (Proc. First Los Alamos Sympos. Math. in Natural Sci., Los Alamos, N.M., 1974), Academic Press, New York, 1976, pp. 259–270. MR0670608 (58 #32342)

Peter D. Lax and Ralph S. Phillips, *Scattering theory for automorphic functions*, Princeton Univ. Press, Princeton, N.J., 1976. Annals of Mathematics Studies, No. 87. MR0562288 (58 #27768)

Peter D. Lax, *On the factorization of matrix-valued functions*, Comm. Pure Appl. Math. **29** (1976), no. 6, 683–688. MR0425663 (54 #13617)

A. Harten, J. M. Hyman, and P. D. Lax, *On finite-difference approximations and entropy conditions for shocks*, Comm. Pure Appl. Math. **29** (1976), no. 3, 297–322. With an appendix by B. Keyfitz. MR0413526 (54 #1640)

Peter D. Lax, *Almost periodic solutions of the KdV equation*, SIAM Rev. **18** (1976), no. 3, 351–375. MR0404889 (53 #8688)

S. Zaidman and Peter D. Lax, *Problems and Solutions: Solutions of Advanced Problems: 5643*, Amer. Math. Monthly **82** (1975), no. 6, 677. MR1537782

Peter D. Lax, *Almost periodic behavior of nonlinear waves*, Advances in Math. **16** (1975), 368–379. MR0425395 (54 #13351)

―――, *Periodic solutions of the KdV equation*, Comm. Pure Appl. Math. **28** (1975), 141–188. MR0369963 (51 #6192)

P. D. Lax and R. S. Phillips, *A scattering theory for automorphic functions*, Séminaire Goulaouic-Schwartz 1973–1974: Équations aux dérivées partielles et analyse fonctionnelle, Exp. No. 23, Centre de Math., École Polytech., Paris, 1974, pp. 7. MR0407626 (53 #11398)

Peter D. Lax, *Periodic solutions of the KdV equations*, Nonlinear wave motion (Proc. AMS-SIAM Summer Sem., Clarkson Coll. Tech., Potsdam, N.Y., 1972), Amer. Math. Soc., Providence, R.I., 1974, pp. 85–96. Lectures in Appl. Math., Vol. 15. MR0344645 (49 #9384)

P. D. Lax and R. S. Phillips, *Scattering theory for dissipative hyperbolic systems*, J. Functional Analysis **14** (1973), 172–235. MR0353016 (50 #5502)

Peter D. Lax, *Hyperbolic systems of conservation laws and the mathematical theory of shock waves*, Society for Industrial and Applied Mathematics, Philadelphia, Pa., 1973. Conference Board of the Mathematical Sciences Regional Conference Series in Applied Mathematics, No. 11. MR0350216 (50 #2709)

―――, *The differentiability of Pólya's function*, Advances in Math. **10** (1973), 456–464. MR0318411 (47 #6958)

Peter D. Lax, Samuel Z. Burstein, and Anneli Lax, *Calculus with applications and computing. Vol. I*, Courant Institute of Mathematical Sciences, New York University, New York, 1972. Notes based on a course given at New York University. MR0354951 (50 #7428)

P. D. Lax and R. S. Phillips, *Scattering theory for the acoustic equation in an even number of space dimensions*, Indiana Univ. Math. J. **22** (1972/73), 101–134. MR0304882 (46 #4014)

Peter D. Lax, *The formation and decay of shock waves*, Amer. Math. Monthly **79** (1972), 227–241. MR0298252 (45 #7304)

Peter D. Lax and Ralph S. Phillips, *On the scattering frequencies of the Laplace operator for exterior domains*, Comm. Pure Appl. Math. **25** (1972), 85–101. MR0296471 (45 #5531)

―――, *Scattering theory*, Rocky Mountain J. Math. **1** (1971), no. 1, 173–223. MR0412636 (54 #758)

Piter D. Laks, Piter D. Laks, Fillips, Ral′f S., and Ral′f S. Fillips, *Teoriya rasseyaniya* (Russian), Izdat. "Mir", Moscow, 1971. With a translation of "Decaying modes for the wave equation in the exterior of an obstacle" (Comm. Pure Appl. Math **22** (1969), 737–787).]; Translated from the English by N. K. Nikol′skiĭ and B. S. Pavlov; Edited by M. Š. Birman. MR397448 (53 #1307)

Peter Lax, *Shock waves and entropy*, Contributions to nonlinear functional analysis (Proc. Sympos., Math. Res. Center, Univ. Wisconsin, Madison, Wis., 1971), Academic Press, New York, 1971, pp. 603–634. MR0393870 (52 #14677)

Peter D. Lax, *The formation and decay of shock waves*, Visiting scholars' lectures (Texas Tech Univ., Lubbock, Tex., 1970/71), Texas Tech Press, Texas Tech Univ., Lubbock, Tex., 1971, pp. 107–139. Math. Ser., No. 9. MR0367471 (51 #3713)

Peter D. Lax and Ralph S. Phillips, *A logarithmic bound on the location of the poles of the scattering matrix*, Arch. Rational Mech. Anal. **40** (1971), 268–280. MR0296534 (45 #5594)

K. O. Friedrichs and P. D. Lax, *Systems of conservation equations with a convex extension*, Proc. Nat. Acad. Sci. U.S.A. **68** (1971), 1686–1688. MR0285799 (44 #3016)

Peter D. Lax and Ralph S. Phillips, *A correction to: "The Paley-Wiener theorem for the Radon transform".*, Comm. Pure Appl. Math. **24** (1971), 279–288. MR0273310 (42 #8190)

Peter D. Lax, *Approximation of meausre preserving transformations*, Comm. Pure Appl. Math. **24** (1971), 133–135. MR0272983 (42 #7864)

Peter D. Lax and Ralph S. Phillips, *The Paley-Wiener theorem for the Radon transform*, Comm. Pure Appl. Math. **23** (1970), 409–424. MR0273309 (42 #8189)

Peter D. Lax, *Invariant functionals of nonlinear equations of evolution*, Proc. Internat. Conf. on Functional Analysis and Related Topics (Tokyo, 1969), Univ. of Tokyo Press, Tokyo, 1970, pp. 240–251. MR0267254 (42 #2156)

Peter D. Lax and Ralph S. Phillips, *The eigenvalues of the Laplace operator for the exterior problem*, Global Analysis (Proc. Sympos. Pure Math., Vol. XVI, Berkeley, Calif., 1968), Amer. Math. Soc., Providence, R.I., 1970, pp. 147–148. MR0265779 (42 #688)

James Glimm and Peter D. Lax, *Decay of solutions of systems of nonlinear hyperbolic conservation laws*, Memoirs of the American Mathematical Society, No. 101, American Mathematical Society, Providence, R.I., 1970. MR0265767 (42 #676)

P. D. Lax and R. S. Phillips, *Purely decaying modes for the wave equation in the exterior of an obstacle*, Proc. Internat. Conf. on Functional Analysis and Related Topics (Tokyo, 1969), Univ. of Tokyo Press, Tokyo, 1970, pp. 11–20. MR0264239 (41 #8835)

———, *Decaying modes for the wave equation in the exterior of an obstacle.*, Comm. Pure Appl. Math. **22** (1969), 737–787. MR0254432 (40 #7641)

Peter D. Lax, *Nonlinear partial differential equations and computing*, SIAM Rev. **11** (1969), 7–19. MR0244597 (39 #5911)

P. D. Lax and R. S. Phillips, *Scattering theory*, Proc. Internat. Congr. Math. (Moscow, 1966), Izdat. "Mir", Moscow, 1968, pp. 542–545. MR0237960 (38 #6237)

Peter D. Lax, *Integrals of nonlinear equations of evolution and solitary waves*, Comm. Pure Appl. Math. **21** (1968), 467–490. MR0235310 (38 #3620)

K. O. Friedrichs and P. D. Lax, *On symmetrizable differential operators*, Singular Integrals (Proc. Sympos. Pure Math., Chicago, Ill., 1966), Amer. Math. Soc., Providence, R.I., 1967, pp. 128–137. MR0239256 (39 #613)

P. D. Lax and L. Nirenberg, *A sharp inequality for pseudo-differential and difference operators.*, Singular Integrals (Proc. Sympos. Pure Math., Chicago, Ill., 1966), Amer. Math. Soc., Providence, R.I., 1967, pp. 213–217. MR0234105 (38 #2424)

P. D. Lax and R. S. Phillips, *Scattering theory for transport phenomena*, Functional Analysis (Proc. Conf., Irvine, Calif., 1966), Academic Press, London; Thompson Book Co., Washington, D.C., 1967, pp. 119–130. MR0220099 (36 #3166)

Peter D. Lax, *Hyperbolic difference equations: A review of the Courant-Friedrichs-Lewy paper in the light of recent developments*, IBM J. Res. Develop. **11** (1967), 235–238. MR0219247 (36 #2330)

P. D. Lax and R. S. Phillips, *The acoustic equation with an indefinite energy form and the Schrödinger equation*, J. Functional Analysis **1** (1967), 37–83. MR0217441 (36 #531)

Peter D. Lax and Ralph S. Phillips, *Scattering theory*, Pure and Applied Mathematics, Vol. 26, Academic Press, New York-London, 1967. MR0217440 (36 #530)

J. Glimm and P. D. Lax, *Decay of solutions of systems of hyperbolic conservation laws*, Bull. Amer. Math. Soc. **73** (1967), 105. MR0204826 (34 #4662)

Peter D. Lax and Ralph S. Phillips, *Analytic properties of the Schrödinger scattering matrix*, Perturbation Theory and its Applications in Quantum Mechanics (Proc. Adv. Sem. Math. Res. Center, U. S. Army, Theoret. Chem. Inst., Univ. of Wisconsin, Madison, Wis., 1965), Wiley, New York, 1966, pp. 243–253. MR0208421 (34 #8231)

P. D. Lax and L. Nirenberg, *On stability for difference schemes: A sharp form of Gårding's inequality*, Comm. Pure Appl. Math. **19** (1966), 473–492. MR0206534 (34 #6352)

Peter D. Lax, *Stability of linear and nonlinear difference schemes*, Numerical Solution of Partial Differential Equations (Proc. Sympos. Univ. Maryland, 1965), Academic Press, New York, 1966, pp. 193–196. MR0203957 (34 #3804)

J. F. Adams, Peter D. Lax, and Ralph S. Phillips, *Correction to "On matrices whose real linear combinations are nonsingular"*, Proc. Amer. Math. Soc. **17** (1966), 945–947. MR0201460 (34 #1344)

P. D. Lax, *Numerical solution of partial differential equations*, Amer. Math. Monthly **72** (1965), no. 2, 74–84. MR0181120 (31 #5349)

J. F. Adams, Peter D. Lax, and Ralph S. Phillips, *On matrices whose real linear combinations are non-singular*, Proc. Amer. Math. Soc. **16** (1965), 318–322. MR0179183 (31 #3432)

K. O. Friedrichs and P. D. Lax, *Boundary value problems for first order operators*, Comm. Pure Appl. Math. **18** (1965), 355–388. MR0174999 (30 #5186)

Peter D. Lax and Burton Wendroff, *Difference schemes for hyperbolic equations with high order of accuracy*, Comm. Pure Appl. Math. **17** (1964), 381–398. MR0170484 (30 #722)

Peter D. Lax and Ralph S. Phillips, *Scattering theory*, Bull. Amer. Math. Soc. **70** (1964), 130–142. MR0167868 (29 #5133)

Peter D. Lax, *Development of singularities of solutions of nonlinear hyperbolic partial differential equations*, J. Mathematical Phys. **5** (1964), 611–613. MR0165243 (29 #2532)

―――――, *Survey of stability of different schemes for solving initial value problems for hyperbolic equations*, Proc. Sympos. Appl. Math., Vol. XV, Amer. Math. Soc., Providence, R.I., 1963, pp. 251–258. MR0160336 (28 #3549)

―――――, *On the regularity of spectral densities* (English, with Russian summary), Teor. Verojatnost. i Primenen. **8** (1963), 337–340. MR0156217 (27 #6146)

―――――, *An inequality for functions of exponential type*, Comm. Pure Appl. Math. **16** (1963), 241–246. MR0155976 (27 #5909)

P. D. Lax, C. S. Morawetz, and R. S. Phillips, *Exponential decay of solutions of the wave equation in the exterior of a star-shaped obstacle*, Comm. Pure Appl. Math. **16** (1963), 477–486. MR0155091 (27 #5033)

Peter D. Lax, *Nonlinear hyperbolic systems of conservation laws*, Nonlinear Problems (Proc. Sympos., Madison, Wis., 1962), Univ. of Wisconsin Press, Madison, Wis., 1963, pp. 3–12. MR0146544 (26 #4066)

Peter D. Lax and Burton Wendroff, *On the stability of difference schemes*, Comm. Pure Appl. Math. **15** (1962), 363–371. MR0154427 (27 #4375)

Peter D. Lax, *A procedure for obtaining upper bounds for the eigenvalues of a Hermitian symmetric operator*, Studies in mathematical analysis and related topics, Stanford Univ. Press, Stanford, Calif, 1962, pp. 199–201. MR0149294 (26 #6784)

Peter D. Lax, Cathleen S. Morawetz, and Ralph S. Phillips, *The exponential decay of solutions of the wave equation in the exterior of a star-shaped obstacle*, Bull. Amer. Math. Soc. **68** (1962), 593–595. MR0142890 (26 #457)

Peter D. Lax and Ralph S. Phillips, *The wave equation in exterior domains*, Bull. Amer. Math. Soc. **68** (1962), 47–49. MR0131059 (24 #A913)

Peter D. Lax, *On the stability of difference approximations to solutions of hyperbolic equations with variable coefficients*, Comm. Pure Appl. Math. **14** (1961), 497–520. MR0145686 (26 #3215)

P. D. Lax, *Translation invariant spaces*, Proc. Internat. Sympos. Linear Spaces (Jerusalem, 1960), Jerusalem Academic Press, Jerusalem; Pergamon, Oxford, 1961, pp. 299–306. MR0140931 (25 #4345)

Peter Lax and Burton Wendroff, *Systems of conservation laws*, Comm. Pure Appl. Math. **13** (1960), 217–237. MR0120774 (22 #11523)

P. D. Lax and R. S. Phillips, *Local boundary conditions for dissipative symmetric linear differential operators*, Comm. Pure Appl. Math. **13** (1960), 427–455. MR0118949 (22 #9718)

Peter D. Lax, *The scope of the energy method*, Bull. Amer. Math. Soc. **66** (1960), 32–35. MR0117443 (22 #8222)

―――――, *Book Review: Numerical analysis and partial differential equations*, Bull. Amer. Math. Soc. **65** (1959), no. 6, 342–343, DOI 10.1090/S0002-9904-1959-10363-3. MR1566018

―――――, *On difference schemes for solving initial value problems for conservation laws*, Symposium on questions of numerical analysis: Proceedings of the Rome Symposium (30 June-1 July 1958) organized by the Provisional International Computation Centre, Libreria Eredi Virgilio Veschi, Rome, 1959, pp. 69–78. MR0107982 (21 #6703)

―――――, *Translation invariant spaces*, Acta Math. **101** (1959), 163–178. MR0105620 (21 #4359)

P. D. Lax, *Differential equations, difference equations and matrix theory*, Comm. Pure Appl. Math. **11** (1958), 175–194. MR0098110 (20 #4572)

Peter D. Lax, *Asymptotic solutions of oscillatory initial value problems*, Duke Math. J. **24** (1957), 627–646. MR0097628 (20 #4096)

P. D. Lax, *A Phragmén-Lindelöf theorem in harmonic analysis and its application to some questions in the theory of elliptic equations*, Comm. Pure Appl. Math. **10** (1957), 361–389. MR0093706 (20 #229)

―――――, *Remarks on the preceding paper*, Comm. Pure Appl. Math. **10** (1957), 617–622. MR0093702 (20 #225)

―――――, *Hyperbolic systems of conservation laws. II*, Comm. Pure Appl. Math. **10** (1957), 537–566. MR0093653 (20 #176)

———, *A stability theorem for solutions of abstract differential equations, and its application to the study of the local behavior of solutions of elliptic equations*, Comm. Pure Appl. Math. **9** (1956), 747–766. MR0086991 (19,281a)

R. Courant and P. D. Lax, *The propagation of discontinuities in wave motion*, Proc. Nat. Acad. Sci. U.S.A. **42** (1956), 872–876. MR0081420 (18,399e)

P. D. Lax and R. D. Richtmyer, *Survey of the stability of linear finite difference equations*, Comm. Pure Appl. Math. **9** (1956), 267–293. MR0079204 (18,48c)

Peter D. Lax, *On Cauchy's problem for hyperbolic equations and the differentiability of solutions of elliptic equations*, Comm. Pure Appl. Math. **8** (1955), 615–633. MR0078558 (17,1212c)

R. Courant and P. Lax, *Cauchy's problem for nonlinear hyperbolic differential equations in two independent variables*, Ann. Mat. Pura Appl. (4) **40** (1955), 161–166. MR0076161 (17,856b)

Peter D. Lax, *Reciprocal extremal problems in function theory*, Comm. Pure Appl. Math. **8** (1955), 437–453. MR0071520 (17,140d)

———, *Symmetrizable linear transformations*, Comm. Pure Appl. Math. **7** (1954), 633–647. MR0068116 (16,832d)

P. D. Lax, *The initial value problem for nonlinear hyperbolic equations in two independent variables*, Contributions to the theory of partial differential equations, Annals of Mathematics Studies, no. 33, Princeton University Press, Princeton, N. J., 1954, pp. 211–229. MR0068093 (16,828b)

P. D. Lax and A. N. Milgram, *Parabolic equations*, Contributions to the theory of partial differential equations, Annals of Mathematics Studies, no. 33, Princeton University Press, Princeton, N. J., 1954, pp. 167–190. MR0067317 (16,709b)

Peter D. Lax, *Weak solutions of nonlinear hyperbolic equations and their numerical computation*, Comm. Pure Appl. Math. **7** (1954), 159–193. MR0066040 (16,524g)

———, *Nonlinear hyperbolic equations*, Comm. Pure Appl. Math. **6** (1953), 231–258. MR0056176 (15,36a)

Paul W. Berg and Peter D. Lax, *Fourth order operators*, Univ. e Politecnico Torino. Rend. Sem. Mat. **11** (1952), 343–358. MR0052661 (14,653c)

Peter D. Lax, *On the existence of Green's function*, Proc. Amer. Math. Soc. **3** (1952), 526–531. MR0051380 (14,470d)

———, *A remark on the method of orthogonal projections*, Comm. Pure Appl. Math. **4** (1951), 457–464. MR0044677 (13,459b)

R. Courant and P. Lax, *Method of characteristics for the solution of nonlinear partial differential equations*, Symposium on theoretical compressible flow, 29 June 1949., Rep. NOLR-1132, Naval Ordnance Laboratory, White Oak, Md., 1950, pp. 61–71. MR0037987 (12,337c)

Peter David Lax, *NON-LINEAR SYSTEMS OF HYPERBOLIC PARTIAL DIFFERENTIAL EQUATIONS IN TWO INDEPENDENT VARIABLES*, ProQuest LLC, Ann Arbor, MI, 1949. Thesis (Ph.D.)–New York University. MR2594102

Richard Courant and Peter Lax, *On nonlinear partial differential equations with two independent variables*, Comm. Pure Appl. Math. **2** (1949), 255–273. MR0033443 (11,441c)

Peter D. Lax, *The quotient of exponential polynomials*, Duke Math. J. **15** (1948), 967–970. MR0029982 (10,693a)

———, *Proof of a conjecture of P. Erdös on the derivative of a polynomial*, Bull. Amer. Math. Soc. **50** (1944), 509–513. MR0010731 (6,61f)

APPENDIX 5

The closed graph theorem

The Lax-Richtmyer equivalence theorem, expounded in Chapter 4, is one of Peter's most famous contributions, and its proof is accessible without specialized preparation. As mentioned there, it is a direct consequence of the Closed Graph Theorem. For readers who have had some graduate-level mathematics, here is the elegant proof of that theorem from pages 168–171 of Peter's *Functional Analysis*.

Theorem 8. *Let X, U, W denote Banach spaces, $\mathbf{M}$ and $\mathbf{N}$ bounded linear maps.*

$$\mathbf{M}: X \to U, \qquad \mathbf{N}: U \to W.$$

Then the composite $\mathbf{NM}$ is a bounded linear map: $X \to W$ with the following properties:

(i) *Submultiplicativity, $|\mathbf{NM}| \leq |\mathbf{N}|\,|\mathbf{M}|$.*
(ii) $(\mathbf{NM})' = \mathbf{M}'\mathbf{N}'$.

Proof. Applying inequality (4) twice, we get

$$|\mathbf{NM}x| \leq |\mathbf{N}|\,|\mathbf{M}x| \leq |\mathbf{N}|\,|\mathbf{M}|\,|x|.$$

Applying definition (3), we get

$$|\mathbf{NM}| = \sup \frac{|\mathbf{NM}x|}{|x|} \leq |\mathbf{N}|\,|\mathbf{M}|. \tag{19}$$

We turn to (ii): applying (8) twice, we get

$$(\mathbf{NM}x, m) = (\mathbf{M}x, \mathbf{N}'m) = (x, \mathbf{M}'\mathbf{N}'m). \tag{20}$$

□

Exercise 8. Prove that multiplication of maps is a continuous operation in the strong topology on the unit balls of $\mathcal{L}(X, U)$ and $\mathcal{L}(U, W)$.

Definition. Two maps $\mathbf{A}$ and $\mathbf{M}$ of a linear space X into itself are said to *commute* if $\mathbf{AM} = \mathbf{MA}$.

Exercise 9. Let X denote a Banach space, $\mathbf{A}$ a bounded map: $X \to X$ that commutes with each of a collection $\{\mathbf{M}_v\}$ of bounded maps $X \to X$. Show that then $\mathbf{A}$ commutes with every map $\mathbf{M}$ that lies in the closed linear span of the set of maps $\{\mathbf{M}_v\}$ in the weak topology.

Exercise 10. Show that in a complex Hilbert space $(\mathbf{NM})^* = \mathbf{M}^*\mathbf{N}^*$.

15.5. THE OPEN MAPPING PRINCIPLE.

The next group of results, the open mapping principle, and the closed graph theorem, goes considerably deeper than the foregoing material. These ideas are due to Stefan Banach; their validity is far from being intuitively clear at first glance, or even a second one.

Theorem 9. *X and U are Banach spaces, and* $\mathbf{M}\colon X \to U$ *a bounded linear mapping of X onto all of U. Then there is a $d > 0$ such that the image of the open unit ball in X under* $\mathbf{M}$ *contains the ball of radius d in U:*

$$\mathbf{M} B_1(0) \supset B_d(0). \tag{21}$$

Proof. Denote by B_n the open ball of radius n around the origin in either the space X or U. Since $\mathbf{M}$ is assumed to map X *onto* U, and since the union of all the B_n is all of X, it follows that $\cup \mathbf{M} B_n = U$. Since the Banach space U is complete, it follows from the Baire category principle that at least one of the sets $\mathbf{M} B_n$ is *dense* in some open set. Some translate of this set is dense in some ball around the origin; since the range of $\mathbf{M}$ is all of U, by linearity of $\mathbf{M}$ we take that translate to be of the form $\mathbf{M}(B_n - x_0)$. The set $B_n - x_0$ is contained in the ball of radius $n + |x_0|$ around the origin. So by homogeneity of $\mathbf{M}$, we conclude that $\mathbf{M} B_1(0)$ is dense in $B_r(0)$ for some $r > 0$. Consequently for any $c > 0$,

$$\mathbf{M} B_c(0) \quad \text{is dense in} \quad B_{cr}(0). \tag{22}$$

We want to show that any point u in $B_r(0)$ is the image of some point x in $B_2(0)$:

$$\mathbf{M} x = u. \tag{23}$$

This point x in $B_2(0)$ is constructed as an infinite series

$$x = \sum_1^\infty x_j. \tag{23'}$$

The terms x_j are constructed recursively: x_1 is taken as a point satisfying

$$|u - \mathbf{M} x_1| < \frac{r}{2}, \qquad |x_1| < 1; \tag{24a}$$

by (22), with $c = 1$, there is such an x_1. We choose x_2 as a point satisfying

$$|u - \mathbf{M} x_1 - \mathbf{M} x_2| < \frac{r}{4}, \qquad |x_2| < \frac{1}{2}; \tag{24b}$$

it follows from (22), with $c = \frac{1}{2}$, and (24a) that such an x_2 exists. Generally, we choose x_m to satisfy

$$\left| u - \sum_1^m \mathbf{M} x_j \right| < \frac{r}{2^m}, \qquad |x_m| < \frac{1}{2^{m-1}}; \tag{24c}$$

it follows from (22), with $c = 1/2^{m-1}$ and (24c) that there is such an x_m.

We noted in chapter 5 on the geometry of normed spaces that if the sum of the norms $\sum |x_j|$ of a series in a complete normed linear space X converges, the series $\sum x_j$ converges strongly. Since by (24c), $|x_j| < 1/2^{j-1}$, it follows that $\sum_1^\infty x_j$ converges to a point x in X, and

$$|x| \le \sum_1^\infty |x_j| < \sum_1^\infty \frac{1}{2^{j-1}} = 2. \tag{25}$$

Since $\mathbf{M}$ is a bounded map, letting $m \to \infty$ in (24c), we conclude that $\mathbf{M} x = \sum_1^\infty \mathbf{M} x_j = u$. $\square$

Theorem 9 has a number of interesting and important consequences; the first one is the *open mapping principle*:

Theorem 10. *X and U are Banach spaces, $\mathbf{M}\colon X \to U$ a bounded linear map onto all of U. Then $\mathbf{M}$ maps open sets onto open sets.*

This is an immediate corollary of theorem 9. □

Theorem 11. *X and U are Banach spaces, $\mathbf{M}\colon X > U$ a bounded linear map that carries X one-to-one onto U. Then the algebraic inverse of $\mathbf{M}$ is a bounded linear map of $U \to X$.*

Proof. It follows from (21) of theorem 9 that for every u in U of norm $d/2$, there is an x in the unit ball of X such that $\mathbf{M}x = u$; note that $|x| \leq 1 = 2|u|/d$. Since $\mathbf{M}$ is homogeneous, it follows that for every u in U there is an x in X, such that
$$\mathbf{M}x = u, \qquad |x| \leq 2|u|/d. \tag{26}$$
Since $\mathbf{M}$ is assumed one-to-one, $x = \mathbf{M}^{-1}u$. Cleary, from (26), $|\mathbf{M}^{-1}| \leq 2/d$. □

Definition. A map $\mathbf{M}\colon X \to U$ from one Banach space into another is called *closed* if whenever $\{x_n\}$ is a sequence in X such that
$$x_n \to x \quad \text{and} \quad \mathbf{M}x_n \to u \tag{27}$$
then
$$\mathbf{M}x = u. \tag{27'}$$

If $\mathbf{M}$ is continuous, it is obviously closed. It is surprising but true that conversely, a closed linear map of a Banach space into another is continuous:

Theorem 12. *X and U are Banach spaces, $\mathbf{M}\colon X \to U$ is a closed linear map.*

Assertion. $\mathbf{M}$ *is continuous.*

Proof. Define the linear space G to consist of all pairs g of form
$$g = \{x, \mathbf{M}x\}, \qquad x \text{ in } X. \tag{28}$$
We define the following norm for g in G:
$$|g| = |x| + |\mathbf{M}x| \tag{28'}$$
Clearly, this is a norm. It follows from (27), (27') and the completeness of X and U that G is complete under this norm. Define the mapping $\mathbf{P}\colon G \to X$ to be the projection onto the first component, that is,
$$g = \{x, \mathbf{M}x\}, \qquad \mathbf{P}g = x. \tag{29}$$
By definition (28') of $|g|$, $|\mathbf{P}g| \leq |g|$, meaning that $\mathbf{P}$ is a bounded operator, $|\mathbf{P}| \leq 1$. Clearly, $\mathbf{P}$ is linear and maps G one-to-one onto X. Therefore, by theorem 11, the inverse of $\mathbf{P}$ is bounded; that is, there is a constant c such that $c|\mathbf{P}g| \geq |g|$. In view of the definition (29) of $\mathbf{P}$ and (28') of $|g|$, it follows that $(c-1)|x| \geq |\mathbf{M}x|$, meaning that $\mathbf{M}$ is bounded. □

The space G defined by (28) is called the *graph* of the mapping $\mathbf{M}$. Requiring $\mathbf{M}$ to be closed is the same as requiring its graph to be closed. Theorem 12 is known as the *closed graph theorem*. The closed graph theorem has many surprising applications.

Theorem 13. *X is a linear space equipped with two norms $|x|_1$ and $|x|_2$ that are compatible in the following sense: If a sequence $\{x_n\}$ converges in both norms, the two limits are equal.*

Suppose that X is complete with respect to both norms; then the two norms are equivalent. That is to say, there is a constant c such that for all x in X, $|x|_1 \leq c|x|_2$, $|x|_2 \leq c|x|_1$.

Proof. Denote by X_1, resp. X_2 the space X under the 1-, resp. 2-norm. By hypothesis, both X_1 and X_2 are complete. Compatibility clearly means that the identity map between X_1 and X_2 is closed. Therefore, by the closed graph theorem, it is bounded in both directions. □

Theorem 14. *X and U are Banach spaces, $\mathbf{M} \colon X \to U$ a bounded linear map. Assume that the range $R_\mathbf{M}$ is a finite-codimensional subspace of U; then $R_\mathbf{M}$ is closed.*

Exercise 11. Prove theorem 14. (Hint: Extend $\mathbf{M}$ to $X \oplus Z$ so that its range is all of U.)

Exercise 12. Show that for every infinite-dimensional Banach space there are linear subspaces of finite codimension that are not closed. (Hint: Use Zorn's lemma.)

Theorem 15. *X is a Banach space, Y and Z closed subspaces of X that complement each other: $X = Y \oplus Z$, in the sense that every x in X can be decomposed uniquely.*

APPENDIX 6

List of Peter Lax's doctoral students (from the Mathematics Genealogy Project)

APPENDIX 6

Name	School	Year	Descendants
Milton Rose	New York University	1953	
Susan Hahn	New York University	1957	
Gideon Peyser	New York University	1957	
George Pimbley	New York University	1957	
Peter Treuenfels	New York University	1957	
Robert Kalaba	New York University	1958	
Burton Wendroff	New York University	1958	
Lucien Neustadt	New York University	1960	
James Moeller	New York University	1961	3
Linus Foy	New York University	1962	
Reuben Hersh	New York University	1963	13
Kennard Reed	New York University	1963	
Norman Rushfield	New York University	1963	
Donald Quarles, Jr.	New York University	1964	
Hector Fattorini	New York University	1965	2
George Logemann	New York University	1965	
Alexandre Chorin	New York University	1966	198
Charles Goldstein	New York University	1967	
Arnold Lapidus	New York University	1967	
James La Vita	New York University	1967	
Julian Prince	New York University	1967	
Li-an Laurence Chen	New York University	1968	
Melvyn Ciment	New York University	1968	
Porter Gerber	New York University	1968	

LIST OF PETER LAX'S DOCTORAL STUDENTS

Name	School	Year	Descendants
Milton Halem	New York University	1968	
Gary Deem	New York University	1969	
Donald Isaac	New York University	1970	
Barbara Keyfitz	New York University	1970	2
Blair Swartz	New York University	1970	
Homer Walker	New York University	1970	9
William Goodhue	New York University	1971	
Gray Jennings	New York University	1971	
Jeffrey Rauch	New York University	1971	28
Steven Alpern	New York University	1973	
Amiram Harten	New York University	1974	
Michael Ghil	New York University	1975	69
Carlos De Moura	New York University	1976	7
James Hyman	New York University	1976	5
Gregory Beylkin	New York University	1982	3
Charles Levermore	New York University	1982	37
Stephanos Venakides	New York University	1982	11
Charles Epstein	New York University	1983	5
Satish Anjilvel	New York University	1984	
Gui-Qiang Chen	Chinese Academy of Sciences	1987	15
Kayyunnapara Joseph	New York University	1987	
Andrew Winkler	New York University	1987	
Yiorgos Smyrlis	New York University	1989	1
Sebastian Noelle	New York University	1990	12

Name	School	Year	Descendants
Spyridon Kamvissis	New York University	1991	
Feiran Tian	New York University	1991	
Yahan Yang	New York University	1991	
Shlomo Engelberg	New York University	1994	
Brian Hayes	New York University	1994	
Min Chen	New York University	1996	
Mikhail Teytel	New York University	1996	

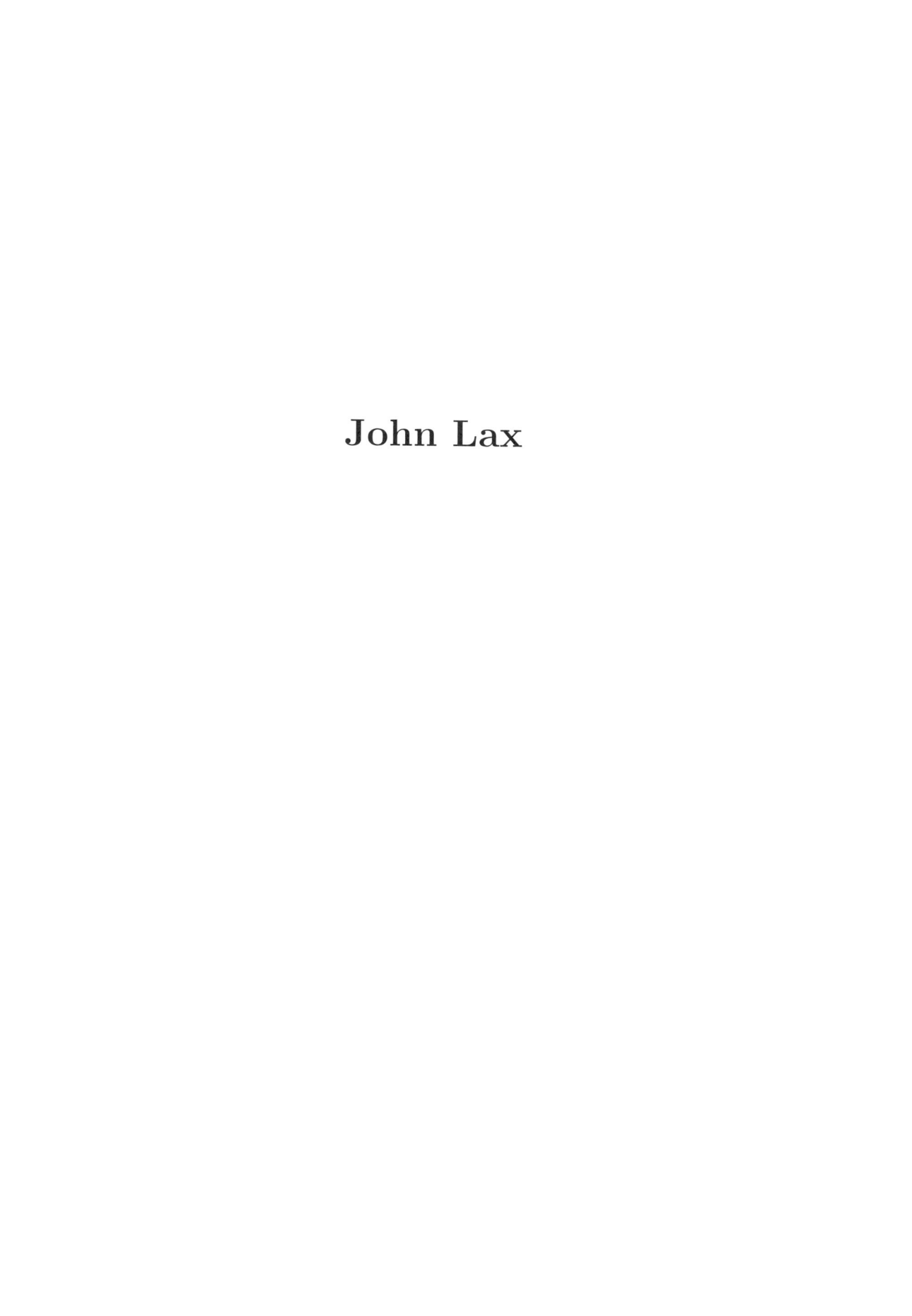

John Lax

I have included two appendices as a memorial to Peter's older son Johnny. The first one is an excerpt from *A Liberal Education*, by John's mentor, the historian Abbott Gleason. The second is John's undergraduate honor's thesis, which appeared in the *Journal of Jazz Studies*.

APPENDIX 7

From *A Liberal Education*, by Abbott Gleason, pages 314–317 on John Lax

Not entirely surprisingly, jazz played a big role in my academic as well as private life. One day in the spring of 1970, a tall, lanky, curly-haired undergraduate with a ready grin dropped by my office, and said he had heard that I was into jazz. He was, too. His name was John Lax, and he turned out to be a most interesting character and an important figure in my life. Jazz was our first bond, but far from our only one. His parents were distinguished mathematicians associated with New York University's Courant Institute. His father, Peter Lax, was and is a world-class mathematician. His mother, Anneli, was a very considerable figure in mathematics pedagogy. Peter Lax was a Hungarian Jew, his wife German. They were refugees from Hitler, and were cosmopolitan Europeans to the core.[†] Their son John, born and raised in the U.S., was not surprisingly less Europe-oriented in both personal style and intellectual interests. His ran to history, literature and music, rather than science. A devoted son, he was nevertheless moving gently away from his parents' world, and in search of something of his own. I think he wanted quite consciously for that "something" to be definingly American. Although John had his share of rueful, ironic humor and could do shtick, his straightforward, extroverted optimism differentiated him somewhat from his parents.

We became close friends almost immediately, having as we did connections to Eastern Europe and black American culture which both dovetailed and contrasted. No doubt this constellation of Jewishness, jazz, history and Eastern Europe linked our friendship psychologically with my old Harvard gang, with whom relations were attenuating at various rates of speed, but remained powerful in my memory. We exchanged music and jokes. We often just hung out. John was a talker, like me, but Sarah minded it less with John, not being married to him. She and I and the kids visited John's family summer place in upstate New York at Loon Lake, where John and I laid out a tennis court together. Our friendship derived some of its power for me from continuity. The old set of Harvard relationships were subject to the pressures of differing careers, interests, regions, to say nothing of family lives. But here was some new growth.

The echoes, and in some cases more than the echoes, of my group of undergraduate friends have remained with me all my life and in some cases grown stronger. Learning how to combine academic and intellectual friendships with emotional relationships, marriage and family began there, in that place. It was a process not unlike learning to reconcile the Edwardian, sentimental side of myself with the aggressive, competitive aspiring wit who also inhabited my skin. Later, as we entered

---

[†]Peter's father (John's Hungarian grandfather) was the doctor who attended Béla Bartok at Mt. Sinai Hospital in his last illness.

John Lax

Photograph by Sarah Gleason. Courtesy of Abbott Gleason, author of A Liberal Education, *and Tide Pool Press, LLC, the publisher.*

old age, some of us began to draw closer again, conscious both of our diverse life trajectories but also of the common, unifying experience of our early friendship and how shaping it had been for us.

In his senior year, John Lax produced a wonderful honors thesis: a profile of the black jazz musician in the Chicago of the 1920s and his experience, discussing the employment situation, the dual black and white union structure, crime, club owners and their clubs, corruption. It was based not only on wide reading in primary and secondary literature but on eight fine oral interviews with survivors of that time and place.‡

After graduating from Brown, John embarked on his Ph.D. in American History at Columbia, and by 1978 was well launched on his dissertation with William Leuchtenburg, a great student of Franklin Roosevelt's world. The subject was to be the American Legion, but in February of 1978, while passing through Chicago on his way to the Herbert Hoover Presidential Library in Iowa, John was killed by a drunk driver. I got the news on a Sunday morning in the kitchen, with my mother

---

‡A portion of John's dissertation was published in the *Journal of Jazz Studies* (June, 1974). The eight interviews are housed in the Oral History Research office (OHRO) at Columbia University. See also Burton W. Pettis, "Speaking in the Groove: Oral History and Jazz," *The Journal of American History*, Vol. 88, No. 2 (September 2001), p. 8.

finally on her deathbed at the farm. I will never forget that bleak, miserable February morning. John was twenty-seven. A wonderful graduate school friend of his, Bill Pencak, now a history professor at Penn State, took over the project and using John's material turned it into a book.[§] Pencak was really the author but John was very much present in it. Had Pencak not resurrected it, John's fine foundation never would have seen the light of day.

John's death devastated his family and friends and produced an aching void in all of us. It is a sadness that I have never entirely gotten over, perhaps because it foreshadowed the inevitable losses to come.

[§]William Pencak, *For God & Country. The American Legion*, 1919-1941. Boston, Northeastern University Press, 1989. See also John Lax and William Pencak, "The Knowles Riot and the Crisis of the 1740s in Massachusetts," *Perspectives in American History*, Vol. X (1976), pp. 163–214.

APPENDIX 8

John Lax article on Chicago jazz musicians

CHICAGO'S BLACK JAZZ MUSICIANS
IN THE TWENTIES: PORTRAIT OF AN ERA

Ever since F. Scott Fitzgerald coined the expression "Jazz Age," historians have used it to conjure up a picture of the unclassifiable cultural and social trends of the 1920s. The label is an appropriate one, if only because jazz did attain widespread commercial popularity and artistic maturity during this decade. By regularly invoking its name as a descriptive device, scholars have implicitly granted that jazz embodied, reflected or proved many of the conditions that contributed to the distinctive atmosphere of the period. Furthermore, as a music developed and played primarily by blacks, yet immensely attractive to many whites, the study of jazz should present opportunities for the social scientist and cultural historian interested in understanding the influences that passed between the races in the postwar decade. Yet jazz—the men who played it, the conditions under which it developed and the effect it had on those who listened and dance to it—has rarely been a topic of serious study.

This essay, based on the testimony of black jazzmen,[1] describes their experience in the 1920s and the response that the new music elicited from other blacks who were exposed to it. An examination of the perceptions of black musicians reveals attributes and conditions, in both black and white society, that have not been clearly identified or understood by more conventional historical analyses.

Some very perceptive work has already been done by Morroe Berger and Neil Leonard on negative reactions to jazz. Focusing on nonmusical values ascribed to jazz by those who resisted its popularity and diffusion, Berger and Leonard uncover, among other things, underlying anti-Negro sentiments in the attitudes of many whites.[2] Their findings also suggest that a powerful generational conflict existed between the youth who embraced the new music and their parents who abhorred the bootleg gin, fast cars, and loose morals associated with it. If middle-aged, middle-class conservatives did not regard jazz as the cause of the distressing behavior of the young, it was a convenient and particularly jarring embodiment of all that was wrong with the new values.

Despite determined opposition, white America came to accept, and in some cases to demand, jazz in nightclubs, at dances and on the radio. Leonard analyzes at length the careers of white band-leaders, notably Paul Whiteman, who reaped great profits by watering jazz down to a level acceptable to most white audiences. In contrast, he describes the feelings that led a relatively small group of white

---

John Lax is in the doctoral program in American history at Columbia University, concentrating on social and cultural history of the late nineteenth and early twentieth centuries.

musicians to embrace jazz for its expressive content and attendant lif-style rather than for its commercial value.[3]

Leonard and Berger offer a picture of white people struggling with the attractions and repulsions they felt towards jazz and the extra-musical values they thought it represented. Strangely enough, there has been less systematic study of black society's reaction to jazz or the role of the Negro musician within this society.[4]

While a fair-sized body of literature exists on black jazz and the musicians who played it, much of it is ahistorical in its approach. A great deal of factual information has been collected about the careers and experiences of black musicians in the 1920s. Jazz scholars, however, have too rarely related these data to the prevailing social conditions, so the valuable research on jazz has had little effect on historical depictions of society, black or white, in the postwar decade.

Chadwick Hansen has taken some tentative steps towards integrating the black musician and his music into a broader societal framework.[5] He contends that black jazz became sweeter and more commercially acceptable to white America by 1930 because of a conscious effort by black musicians to absorb white cultural values. This assertion, based on limited evidence and questionable assumptions, will be discussed in greater detail later. Regardless of the validity of its conclusions, Hansen's work is a distinctive effort to examine the feelings and motivations of black musicians from a perspective that encompasses far more than their own social milieu.

It is my contention that jazz musicians, by the middle and late 1920s, were prestigious members of black society. Their influence, though not institutionalized, was proportionally greater than their numbers would suggest. Hansen's article, based largely on the testimony of Dave Peyton, a black band leader in Chicago,[6] is a logical starting point for further discussion of the attitudes of Chicago's black musicians and their influence on the rest of black society during the 1920s.

There were, as Hansen contends, black musicians who were drawn to any music with a soupy, symphonic sound because it conjured up a superficial similarity to a European musical tradition that they imagined was consistent with white values. An examination of the experience of Chicago's black musicians indicates, however, that by the mid-1920s those who unreasoningly advocated the maintenance of classical aesthetic standards no longer accurately represented the sentiments of the majority of their group. Increasingly, the black musicians who became popular as soloists and band leaders were those who had developed a style of their own that was unencumbered by many of the expressive goals and techniques associated with the European tradition but alien to jazz. Indeed, the growing attraction and power of a hot brand of jazz that came to Chicago in the first decades of the century dictated a change in musical tastes. It presented an indigenous musical alternative to musicians and black audiences alike. During this period of rapid development jazz became a tremendously powerful force in its own right. That this new style served to heighten black cultural awareness and establish new standards for artistic achievement will be demonstrated in the ensuing discussion.

Before focusing on Chicago in the 1920s, it is necessary to briefly outline the plight of black musicians in the South and North before jazz and blues attained great popularity. When compared with the period preceding it (with some overlapping) the 1920s represent a period of cultural awakening in black society and increasing status for the musician within this society.

II

During the pre-World War I days when minstrelsy was still in fashion, the travelling Negro show people and musicians met few Southerners, black or white, who wanted to have anything to do with them. The eminently respectable W.C. Handy, who held numerous positions as a musical educator as well as show trumpeter, discovered "That minstrels were a disreputable lot in the eyes of a large section of upper crust Negroes."[7] Pops Foster, a New Orleans musician who travelled a good deal in the South, experienced even harsher treatment than Handy as a result of his profession. He found that "musicians were nowhere in the South" and that among a large segment of the black population "show people were classified as nothin' and musicians were rotten."[8]

The stigma that entertainers bore among Southern Negroes characterized a provincialism shared by most of rural America. During the same period, Northern, urbanized blacks also demonstrated a selective distaste for certain forms of black music and those who played it. Garvin Bushell observed that New York's more established Negro families wished to suppress the blues.

> Most of the Negro population in New York had either been born there or had been in the city so long they were fully acclimated. They were trying to forget the traditions of the South; they were trying to emulate the whites.... You usually weren't allowed to play blues and boogie woogie in the average Negro middle-class home. The music supposedly suggested a low element.... You could only hear the blues and real jazz in the gutbucket cabarets where the lower class went.[9]

By 1920, blues and jazz had begun to gain popularity in Harlem, despite the continued resistance of certain elements within that community.

Blues singer and songwriter Perry Bradford claims to have organized one of the earliest blues revues presented in Harlem. He felt that in the decade preceding the 1920s the only impediment to the popularity of blues in Harlem lay in the pretensions of a coterie of upper class blacks.

> It was confusing to see some of those "Hate Blues" hypocrites, who were preaching and brain-washing before the public how much they detested the blues, yet whenever the same so-called sophisticated intellectuals and top musicians would hear some low-down blues sung and played at a House-Rent Party or some hole in the wall speakeasy, they'd let their hair down, act their age, be themselves, and go to town belly-rubbing and shouting, "Play 'em daddy—if it's all nite long."[10]

In 1920, the OKEH record company released the first records made by a black blues singer. The unprecedented sales in black urban centers of "Crazy Blues" and "It's Right Here for You," sung by Mamie Smith and her Jazz Hounds, proved the existence of a large number of Negroes willing to buy down-home style singing, recorded by a member of their own race. Bradford, composer of these two songs and the force behind this major commercial breakthrough in the recording industry, triumphantly described its reception in 1920 by the black community of Chicago:

> Since colored folks had never before heard their blues and rhythms recorded by one of their own... when Mamie came to Chicago, on

the Theater Owners and Booking Association (T.O.B.A., nicknamed Tough on Black Artists) Circuit, it was a spectacular scene, for she brought out a shoving crowd two blocks long.... Clarence Williams had three music stores in Chicago at the time and Richard M. Jones, Williams' general manager of the downtown store at State and Lake Streets, told me that they couldn't supply the demand for Mamie's first record.... which came out under the OKEH label and was selling at one buck per platter. The demand was so great that many people left two-dollar deposits to be sure of getting one. Pullman porters bought them by the dozens at a dollar per copy and sold them in rural districts for two dollars.[11]

Three years later, pianist Willie "the Lion" Smith took note of the great change in popular tastes: "In the old days, uptown or downtown, they would often chase you away from the piano when you played the low-down blues. The time was coming that if you couldn't play the blues and get in stride, you couldn't find work."[12]

Blues singing had been popular for some time in the travelling shows of various kinds, but in 1920 it widened its appeal and profitability. Blues issued for the Negro market on so-called "race records" found great popularity in the South and in a number of Northern cities, most specifically Chicago. Similarly, instrumental jazz musicians who were strongly influenced by the blues (as were many who came from the South) increasingly migrated towards a Chicago market clamoring for authentic, Southern-style jazz and blues. Both Southern and native born black musicians encountered a new and for the most part inviting environment in Chicago.

Chicago received a great influx of Southern blues-oriented jazz musicians well before New York. In the beginning, the big theater bands were "composed of all legitimate musicians"[13] rather than jazz improvisors. Many club owners, including Bill Bottoms of Dreamland, said they "didn't like the idea of a 'jazz' band"[14] in the cabarets. The club owners, however, soon realized that they had to feature a jazz band if they wished to remain in business. By the 1920s in Chicago as well as New York, jazz musicians, some of them strongly rooted in the blues, had attained commercial acceptance, economic prosperity, and social prestige within their communities. In the words of Chicago clarinetist Scoville Browne:

> You were like Mr. Sousa. You were something in other words. You were quite a boy.... It was a good profession in those days. It was a white collar thing. It was an artistic thing. You were in a high income bracket. There wasn't too many of those people around unless they were professionals, doctors or lawyers or something like that. They weren't too plentiful either.

While blues-oriented jazzmen and hot improvisors gained increasing prestige and success during the 1920s, they were not the only element in black Chicago's musical hierarchy. They encountered a structured musical world largely controlled by more conventional and less inspired men who nevertheless knew how to make music pay.

Within the increasingly complex structure of black Chicago's music business there developed a demand for musicians of widely varying backgrounds, skills, and abilities. The silent movie houses required pit orchestras to accompany the feature and to give concerts between shows; vaudeville theaters needed house bands to support the ever changing acts on the T.O.B.A. circuit; ballrooms and dance halls called for bands that could play dance music; and cabarets required smaller ensembles to play for dancers, to support floor shows, and to improvise in a swinging style. Private orchestras hired out to play society affairs, often for masonic lodges and other exclusive organizations. At the other end of the social spectrum itinerant blues musicians, guitarists and pianists, entertained at bars all over the South Side. A musician needed a good deal of versatility to play successfully in several of these musical contexts.

As opportunities and musicians multiplied during the first quarter of the century, several musicians who had established themselves in Chicago well before 1920 seemed to exert some commercial and organizational leadership in the musical branch of the black community. First, these emergent leaders had the prestige of leading large and successful vaudeville, dance, or pit orchestras (often 15 or 16 pieces). Second, some gained a reputation and a firm financial foothold as booking agents or as non-playing musical directors and organizers. Third, a significant number attained influence and high office in the only formal instrument of control and organization in their profession—Local 208, the black arm of the Chicago branch of the American Federation of Musicians. Finally, most of them had a high degree of formal musical training and often functioned successfully as instructors or arrangers. Most of these men held aesthetic ideals primarily derived from the European classical tradition, which they tried to incorporate into their music. Like all popular musicians, they had to remain constantly aware of the changing tastes of their public, which, in most instances, was composed mainly of Negroes living on the South Side.

Erskine Tate studied at Lane College in Tennessee and the American Conservatory in Chicago before leading the pit orchestra at the Vendome Theater on Thirty-first and State Streets from 1918 to 1927.[15] This theater represented the height of elegant family entertainment for the black community. Tate's band accompanied first-run silent films and then provided an hour-long concert, ranging from lush classical overatures to hot jazz.

From 1916 to 1922, Charlie Elgar's band played winters at Harmon's Dreamland, a white-owned North Side dance hall and summers at the municipal pier. Originally from New Orleans, Elgar came to Chicago in 1913 and later studied at Coleridge Taylor School of Music. He also attended Marquette University in 1923 or 1924 when he had a band in Milwaukee.[16] He acted as a booker in addition to leading ensembles in his early days in Chicago; he later served for many years as president of Local 208 and was a part-time teacher throughout his early career.[17]

Charles L. "Doc" Cooke came to Chicago from Louisville, Kentucky. After resigning as musical director of Riverview Park, he led his own sixteen-piece band at Harmon's Dreamland from 1922 to 1926 and then had a dance band at the White City Ballroom until 1930. Cooke did a good deal of arranging and achieved the degree of Doctor of Music in 1926 at the Chicago College of Music.[18]

Dave Peyton was born in Louisiana in 1885, but he always considered himself a Chicago musician. In 1908 he held a job at the Little Grand Theater on Thirty-first Street and State playing piano for Wilbur Sweatman, a touring vaudeville clarinetist with a vigorous ragtime style. With the construction, in March 1911, of the Grand Theater next door to the Little Grand, Peyton became the leader of a vaudeville trio.[19] In the 1920s, Peyton pursued a dazzling array of activities, not all of them musical. He acted as a band contractor, supplying orchestras for specific engagements. He wrote the music for the musical "That Gets It."[20] He worked as an arranger, published his weekly column in *The Chicago Defender* (read, he claimed, by 30,000 musicians),[21] and served on the board of directors of the black musicians' union. In a playing and conducting capacity, he opened at the new Peerless Theater in 1926; two years later he became the first director of the pit orchestra at the mammoth new Regal Theater.[22] In addition, his orchestra played numerous social, municipal, and commercial cabaret engagements throughout the decade.

Since Peyton is one of the few black musicians who left written testimony of that period, it is tempting to regard his column as representative of the feelings of Chicago's black musicians. As later evidence will demonstrate this was clearly *not* the case. However, it is highly plausible that Peyton's opinions did represent the musical organizers and band leaders who reached their zenith during the mid-twenties and who based their success on commercial and business instincts rather than on musical creativity or instrumental virtuosity.

It seems unlikely that Dave Peyton would have been allowed to express his views in print week after week if they had been at great variance with the opinions of the rest of the black musical establishment. He was, at least superficially, the prototype of the successful black band leader of the early 1920s. As the bassist Milt Hinton suggests, Peyton tried to distinguish himself from the less educated musicians who played without benefit of written music at the local cabarets: "He was a highbrow, you know. He was a great arranger and I can remember seeing him—a short dark man, East Indian sort of black color, little hawk nose. I can just see him. He had such dignity and he was the black bourgeois musician."

As the popularity and wealth of the musical profession in Chicago increased, many South Side youngsters began to study and pursue music as a career. They developed a hot style which the older Chicago-born musicians did not have. Clarinetists Darnell Howard and Scoville Browne, pianist Cassino Simpson, trumpeter Ray Nance, saxophonists Leon Washington and Willie Randall, bassist Milton Hinton—these men and many more grew up in Chicago and benefited from the variety of musical styles surrounding them in their formative years. Others, notably Quinn Wilson, Wallace Bishop and Hayes Alvis, gained valuable early experience under Major N. Clark Smith in *The Chicago Defender* Boys' Band. With much greater educational opportunities than the South would have afforded them, they often became technically proficient on their instruments, good readers, and excellent improvisors. Some of them effectively combined the formal musical training of the educated band leaders with the hot style of the less schooled musicians.

During the 1920s the musicians had little direct social contact with the rest of black society. Few members of the working class had the time, money, or inclination to frequent the cafes on the South Side, and the black middle class, as Milt Hinton suggests, preferred not to associate with musicians at all. It acknowledged the existence of black jazz musicians only when the latter started gaining national acclaim and prestige.

> I don't think we were looked on with so much favor. I don't think they even paid attention to us. The black bourgeoisie, as we would prefer to call it were so entrenched with their own

thing and being a special breed that they had come from parents that could afford to give them an education—they had their society so set up, that I don't think they even looked our way until the stars began to show up—the Duke Ellingtons, and the Cab Calloways.... The black bourgeoisie—the doctors and the lawyers, they were just as far apart from us as night and day. I don't remember them ever being a part of us or we being anything more than a service to them as an entertainer whenever they gave a party or a formal affair.

Even when jazzmen had gained status and respect in their community, William Everett Samuels recalled that they were "kind of clannish" and tended to "stay to themselves"; for the most part "musicians hung out with musicians" and did not try to become part of the established black middle class.

There were a few musicians, for example Dave Peyton, who wanted music to join the ranks of middle-class professions, such as law or medicine. He felt that musicians should follow all the conventional standards of proper behavior and that this would lead them to financial success and respectability in the eyes of both the black and white communities. He had a whole list of hints for guiding black musicians along the road to what he called "success." In making these suggestions (don't drink on the job, don't tap your foot while playing, always be neat and clean, etc.), his tone had the one-dimensional preaching ring often associated with Booker T. Washington. Peyton's preoccupation with middle-class, white standards of respectability and cultural achievement made it psychologically impossible for him to be a musical innovator in the jazz idiom. In any case, he considered jazz a degrading music whose only merit was its popularity with a temporarily insane public.

Even from a musical point of view, Peyton never expressed any real interest in jazz. He regarded hot jazz as an amalgam of technical eccentricities—a series of moans and squawks having no artistic meaning. He admired men like Louis Armstrong and Joe Oliver not for their artistic achievement but for their commercial success. He wrote, "Musicians should regard playing music as a business—regard it just as a doctor, lawyer, mechanic and all other tradesmen do their profession. It is your living: so why regard it otherwise."[23] Even in praising the music that he professed to love—classical music and spirituals—he betrayed a lack of understanding and genuine appreciation. Quite obviously, he was attracted by the values he felt these forms embodied, not their essence. Considering his deprecation of jazz, his outspoken admiration for financial success, his desire for prestige, and his emphasis on the maintenance of what he conceived to be professional standards of behavior, Peyton could very well have chosen the musical profession simply because it represented one of the quickest avenues of advancement for an ambitious, capable black man in early twentieth-century urban America.

Until the mid-1920s, Peyton and others with similar values exercised a great deal of control within black Chicago's musical profession. The influx of large numbers of Southern jazzmen, some of whom were self-taught and could not read music, caused confusion in Chicago's musical establishment. It would be misleading to say that the migrants emerged victorious in the battle for control of the music profession in Chicago, because most of them did not seek to become orchestra leaders high-ranking union officials, or bookers. (Hines and Armstrong became orchestra leaders because of their sheer instrumental brilliance, not their organizational talents.) That is, they did not have the same goals as Peyton, Elgar, Tate, Cooke and other influential leaders. Consequently, no direct conflict ensued. However, their music had a tremendous impact on Chicago's black population, while many of the older leaders began to seem out of date, both musically and culturally.

As hot musicians began to infiltrate the pit orchestras of the silent movie houses, a sizeable number of Chicago's black population had their first opportunity to hear swinging jazz and to compare it with the symphonic overtures that had preceded it. It soon became obvious that jazz had overtaken semi-classical and ragtime music in popularity, especially among the young. Milt Hinton, who witnessed this process as an impressionable youngster in Chicago's silent movie theaters during the early 1920s expressed the implications that it had for black society.

> Here we had come up in a society in Chicago from, as I say, the twenties, where we were emulating, if you would say it, the white studio orchestra in the pit playing overtures. But then in our own ethnic way, after the overture was over, there would be a big trumpet solo by Louis Armstrong on "Saint Louis Blues" and going into one of these fantastic things which was quite creative.
>
> But the people would come on Sundays to the theater and they would be dressed. They would have on their tuxedoes with wing collars and it was like we were emulating white folks, like it was a big white theater you know. Then the orchestra played this little overture and then all of a sudden, we'd go right straight back into our own thing.
>
> Black people wanted to be like white people because they felt this was the way to be; that you were right and you were white and this was the only way it could be.... We emulated white people because this was a very conditioned thing that had been brought down to us that this was the only way of life.
>
> Louis had enough of the academic thing to read the music properly, and so this was the style. We were going to be just like downtown. And we'd sit there, my mother would have me by the hand, and we'd sit and listen to this overture which had a European environment. Then the people would be a little restless, and say "Well, that sounds nice," and applaud it. Then somebody would say "Hey baby, play so and so," and when Louis stood up and played one of his great solos, you could see everybody letting their hair down and say "Yeah, tha's the way it should be. This is it." So we were beginning to say "Well, it's great to be like that, but this is what really relates to us." And then the soloists like Stomp Evans, Barney Bigard, Teddy Weatherford, he was a great pianist, and Eddie South would play. He could play the most beautiful melody in the world like any gypsy violin player, and then go into that gut bucket. They really changed things because they had the know-how and the catalysts in order to do these things.

Perhaps one blast from the trumpet of Louis Armstrong was not enough to free instantly all the blacks from the shackles of white cultural bondage, but jazz did more to change the perspective of blacks towards their own heritage than any other cultural movement of the decade.

Black musical innovators managed to combine commercial success with an original aesthetic, thus softening black middle-class opposition to jazz. For even if Peyton and other blacks with middle-class aspirations did not like the way Armstrong played, they recognized and admired his national prominence and financial success. After all, whites dominated classical music more thoroughly than blacks did jazz, though for different reasons. The middle class, who were highly conscious of success of any kind, could not help but realize that if blacks were to rise in the musical profession, it would have to be at least in part within the realm of jazz and popular music.

The self-conscious black intellectuals of the Harlem Renaissance, in contrast with the jazz musicians, set about very deliberately to create their own artistic aesthetics. Unfortunately, they achieved far less than the musicians in both artistic and commercial terms. Nathan Huggins suggests in *The Harlem Renaissance* that they could not free themselves from the traditions of American subservience to European high culture, nor did their work appeal to the greater number of blacks within their own community. Yet, as Huggins goes on to show, they were not aware that jazz musicians were creating a music far less subservient to white standards than anything these intellectuals had created:

> Harlem intellectuals promoted Negro art, but one thing is very curious, except for Langston Hughes, none of them took jazz— the new music—very seriously. Of course, they all mentioned it as background, as descriptive of Harlem life. All said it was important in the definition of the New Negro. But none thought enough about it to try and figure out what was happening. They tended to view it as folk art—the unrefined source for the new art...the promoters of the Harlem Renaissance were so fixed on a vision of high culture that they did not look very hard or well at jazz.[24]

In explaining the failure of an indigenous ethnic theater, Huggins again puts into perspective the achievements of jazz during that period. The lucrative lure of minstrelsy channeled all the black talent into a barren, uncongenial setting for the development of black drama. Created by whites, minstrel shows served deeply rooted psychological needs of whites by depicting blacks as lazy, shiftless, and incompetent. These traits represented the milder vices whites could not admit in themselves. During a minstrel show whites could spend an hour or two of vicarious release while watching Negroes manifest qualities stifled in white society. Were black performers to deviate significantly from the stereotyped roles, the performances would not be well received by white audiences. Since this was where the real money lay, black entertainers directed their creative urge towards accommodating a narrow, stunted form of entertainment.

In contrast to black writers and actors, black musicians found themselves within a tradition of their own creation. Consequently they were at liberty to experiment musically with a freedom that is vital to any growing art form. Moreover their music was favorably received by a sizeable number of both blacks and whites and flourished. It eventually came to be adapted by the music industry for presentation on the growing mass media, not as a recognition of its artistic value but as a tribute to its power to capture the public fancy. Contrary to Hansen's depiction of jazz as a music embodying the hated past (from which Negroes fled headlong into the arms of white commercialism), it proved to be an art form of such expressive power that even white society eventually succumbed. Certainly many blacks gained a more positive sense of racial identity both by hearing the music itself and by identifying with the success of the men of their race who created it.

III

The conviction that by 1930 jazz became hopelessly compromised by its incorporation into American popular music does not fully take into account the great jazz that survived nor the new jazz created since. To be sure, the cabarets, night clubs, and theaters in Chicago had been almost completely wiped out but there was less music of all kinds in the United States. The thirties have often been called the era of the travelling bands, and some of these bands contained fine black musicians. Orchestras like those led by Fletcher Henderson, Duke Ellington, and Earl Hines were still presenting a swinging, improvisational jazz that had been adapted to the big band format without compromising its essence.

Starting in the late 1920s, Kansas City produced a whole contingent of midwestern, blues-oriented Negro musicians who, as Ross Russel points out, represented "a veritable who's who of southwestern musicians concentrated in a single band, or even a section of a band."[25] He also observes that Kansas City, by being far removed from the center of the music industry "escaped the disturbing commercial pressures that began to nag away at jazz toward the end of the Chicago period and were a constant harassment to creativity in New York."[26] The smoothly functioning machine of Boss Prendergast kept Kansas City night life flourishing throughout the Depression years and avoided the tremors that occurred elsewhere as a result of prohibition repeal. Band leader Andy Kirk vividly recalled the striking difference between Kansas City and other towns in the 1930s:

> We went on tour through Arkansas and Oklahoma for the Malco Theater chain. They had a great many houses around the Southwest but nobody had any money to get into the theaters with. That was just around the time Roosevelt called in the gold, right in the middle of the Depression. We gave a final concert in Memphis...and just did get back to Kansas City. When we got back home, there was no Depression. The town was jumping! We got back Friday night and the following Monday went into the Vanity Fair Club, a plush spot right in the center of town and did good business.... We'd get the finest acts out of Chicago to play in the nightclubs in Kansas City because they weren't working regularly.[27]

The example of Kansas City supports the notion that the increase in sweet jazz of the insipid variety can be attributed to two sets of conditions: (1) the growing influence of commercialism that inevitably arose from the technological development of mass media entertainment, and (2) the conditions of the Depression that dried up the market for local musicians, who had hitherto provided nightclub entertainment in cities like Chicago.

For ecomonic reasons, the early thirties did not represent a high point of black musical activity. It was not an era where black musicians consciously tried to exchange their indigenously created musical tradition for an insipid melting pot of

American commercialism; on the other hand, they did want to take an active part in music as a business and reap its economic rewards. There is no reason to view this as any kind of a compromise of artistic integrity. Poverty may have impelled some black musicians to appeal to a larger white audience by adapting certain techniques of a sentimental, vacuous musical style. However, they did not do so because they either wanted to be like white men or because they wanted to be socially accepted by whites. In most cases their sole motive was the desire to make a living. Had the music industry in America functioned as it should have and given blacks an equal opportunity to present their music to the public in an unadulterated form, perhaps some of these musical compromises could have been avoided.[28]

Some rebellious white intellectuals adhered to jazz as the negation of the American middle-class values they detested. William Leuchtenburg noted when surveying the mood of American intellectuals in the 1920s:

> In fear and dislike of a machine age, many intellectuals turned towards more primitive societies, fleeing to Mexico or studying the art of the Congo or centering on the Negro as the symbol of pre-industrial man, uninhibited in his laughter and his sadness. In both the United States and Europe this took the form of a cult of jazz, for in a world of synthetic songs mechanically contrived on Tin Pan Alley, the rhythms of New Orleans and the Delta country had the ring of spontaneity.[29]

They viewed any cooperation between jazz and commercial music as a victory for the oppressive forces of traditional American values. Just as in recent years LeRoi Jones (Imamu Baraka) has seen black music as "always radical in the context of American culture,"[30] certain white jazz enthusiasts also wished to keep the Negro and his art separate from the mainstream of American life in the hope that blacks would, if isolated from American social and cultural norms, supply them with alternative life-styles and values and thus help them stave off disillusionment.

For their part, black jazzmen had no reason to keep their music free from the taint of white influence. From its inception, jazz had combined a variety of African, European, and American features in creating a new form of musical expression. Indeed, its vitality resulted from cultural assimilation and synthesis. If it was to grow, it had to continually draw on all resources in the society around it. The interest in classical music which some Negro jazzmen of the day expressed showed a healthy desire to broaden their horizons and in no way diminished their stature as jazz artists. Only in the case of Peyton and those who shared his uncritical subservience to white, European aesthetic ideals did the acceptance of classical music presage a rejection of the supposed black values embodied by jazz. In any case, jazz transcended the label of protest music, or racial music, and the sociological pigeon hole into which some have sought to put it. As Ralph Ellison has pointed out burdening the music with artificial social values is enough to give "even the blues the blues."[31] On the whole, black jazzmen's concern with European musical ideals stemmed from their intrinsic musical merit.[32]

Hansen's argument that the urge towards respectability and away from a "degraded past" motivated black jazz musicians to emasculate their music by sweetening it with the cloying confection of popular white music was based on the dubious assertion that black jazz was swallowed up in sweet music for a distinct period around 1930. Actually, most jazz musicians concerned themselves a good deal less with social values, white or black, than did their counterpart white musicians. Therefore the implication of social and cultural dependency inherent in the contention that black musicians responded in an accommodating manner to white social norms could be turned around inasmuch as whites seemed a good deal more preoccupied with black music—they plagiarized from it, they were threatened by it, they were drawn to it—than vice versa. I suspect that the origin of Hansen's misinterpretation was his acceptance of Peyton's views as typical of black musicians. In reality they represented only a small group with special interests and values. Rather than acting as leaders of an assimilationist movement to sacrifice the guts of jazz on the altar of white commercialism and middle-class sentiment, Chicago's black musicians in the twenties succeeded in turning the attention of many Northern blacks away from white cultural standards and back to the more indigenous forms of expression that many thought they had left behind them.

The Chicago experience demonstrates the tremendous change that hot, improvisational jazz exerted on black cultural awareness. Before the great New Orleans musicians, as well as those from numerous other localities, came to Chicago, pit orchestras played semi-classical, ragtime, and symphonic music. White society set the aesthetic standards. The orchestras were led by men like Peyton, who stressed the importance of "proper" musical technique and classical training. They understandably felt that a firm maintenance of musical standards was essential if black musicians were to successfully compete with whites in the American market for commercial music. Yet, the instrumental brilliance and musical creativity of the jazzmen wrought a tremendous change in the attitudes of their audiences and in the position of the organized black leadership.

By the 1920s, black audiences grew increasingly restless in their seats during the classical overtures. They were eagerly waiting for Armstrong to begin one of his celebrated hot choruses. If they wished to retain their influence, the orchestra conductors had no choice but to go along with this trend and to give the public what it wanted. Meanwhile, the cultural leadership had passed to a new group of black musicians who proved that hot jazz could be lucrative as well as artistically satisfying.

IV

Technical and stylistic innovations in jazz were spawned by a need for self-expression that simply could not be wholly accommodated within the musical traditions of another age and another culture. The precise mixture of African, European, and American musical components that led to the development of jazz can never be exactly determined; nor can we ever fully know the emotions that led certain men to create jazz and others to accept it. However, to regard it merely as a black reaction to the fact of white oppression is to reduce and distort its meaning. In reality jazz represented a far more profound expression that came not only from the core of black society but also from the core of American society. It was not only the product of the black experience but also of the American experience. It may have been this very quality that most disturbed those whites who opposed it. For if jazz, a music created by blacks, appealed to a widely based group of Americans and expressed the mood of the age, then perhaps the Negro's presence in society exerted a more dominant influence than many wished to acknowledge. He provided the rhythm and melody to which America in the 1920s danced, willingly or unwillingly.

The development of black society has too often been viewed as a reaction to conditions imposed on it from above rather than as a more independently determined process. To be sure, the racial prejudice that permeated America also limited the range of choices available to blacks. Nevertheless, evidence like the birth and development of jazz indicates that forces emanating from black society often influenced whites as well as the other way around. For black society did have many components that originated within itself and did not result directly from either emulating white society or from following its commands. The Afro-American community remained distinct, although not necessarily aloof, from white society, rather than dependent on it.

Scholars acknowledge that the analysis of literary sources yields a distorted picture of the true feelings and trends within a society. This is particularly true in the case of American blacks where the oral and musical tradition was more central to the culture than writing. The use of music and the testimony of musicians as a source of historical and sociological analysis presents methodological problems that must be handled with care. Personal interviews are necessary because only a small and unrepresentative contingent of black jazz musicians have written about their experiences and feelings. On the other hand, to characterize the 1920s as "The Jazz Age" without examining the music and the effect it had on those who heard it is to close off a source of valuable information about relations between blacks and whites. Jazz was a music created by blacks and they supplied the continuing impetus for its development. Although black musicians had to pay attention to white tastes for economic reasons, they set their own artistic and musical standards. That they were the leaders within this creative process is incontrovertible.

Notes

[1] The author interviewed the following eight black musicians who were professionally active in Chicago during the 1920s: Ralph E. Brown, Chicago, Ill., December 29, 1971; Scoville Browne, New York City, December 4, 1971; Earl Hines, New York City, December 1, 1971; Milton Hinton, St. Albans, New York, December 3, 1971; Willie Randall, Chicago, Ill., December 28, 1971; William Everett Samuels, Chicago, Ill., December 29, 1971; Red Saunders, Chicago, Ill., December 24, 1971; Leon Washington, Chicago, Ill., December 27, 1971. All unfootnoted quotations are from these interviews.
[2] Neil Leonard, *Jazz and the White Americans* (Chicago: University of Chicago Press, 1962) and Morroe Berger, "Jazz: Resistance to the Diffusion of a Culture Pattern," *The Journal of Negro History* 32 (Oct. 1947), 461–95; reprinted in Charles Nanry, ed., *American Music: From Storyville to Woodstock* (New Brunswick, N.J.: Transaction, 1972), pp. 11–43. Berger stresses that certain middle-class blacks were also opposed to jazz.
[3] Leonard, *Jazz and the White Americans*, pp. 55–72.
[4] For a provocative, impressionistic historical account of the experience of the black musician and a perceptive discussion of the meaning of jazz and blues within the context of American society and culture see Leroi Jones (Imamu Baraka), *Blues People* (New York: William Morrow, Inc., 1963).
[5] Chadwick Hansen, "Social Influences on Jazz Style: Chicago 1920–1930," *American Quarterly* 12 (Winter 1960): 493–507.
[6] Peyton's column, "The Musical Bunch," appeared weekly in *The Chicago Defender* from October 17, 1925 to August 24, 1929.
[7] William C. Handy, *Father of the Blues* (New York: Macmillan, 1941), p. 33.
[8] Pops Foster, *The Autobiography of a New Orleans Jazzman*, as told to Tom Stoddard, interchapters by Ross Russel (Berkeley and Los Angeles: University of California Press, 1971), p. 69.
[9] Nat Hentoff, "Jazz in the Twenties: Garvin Bushell," in Martin Williams, ed., *Jazz Panorama* (New York: Collier, 1964), p. 74.
[10] Perry Bradford, *Born with the Blues* (New York: Oak Publications, 1965), p. 97.
[11] Ibid., pp. 47–8.
[12] Willie, "The Lion" Smith with George Hoefer, *Music on My Mind* (New York: Doubleday, 1964), p. 129.
[13] Nat Hentoff and Nat Shapiro, *Hear Me Talkin' to Ya* (New York: Dover, 1965), pp. 78–79.
[14] Ibid., p. 85.
[15] John Chilton, *Who's Who of Jazz: Storyville to Swing Street* (London: Bloomsbury Book Shop, 1970), p. 395.
[16] Letter from John Steiner, Chicago, Ill., September 13, 1972. I wish to thank Mr. Steiner, a noted historian of Chicago jazz, for sharing his extensive knowledge with me.
[17] Chilton, *Who's Who*, pp. 123–24.
[18] Ibid., p. 88.
[19] Private conversation with John Steiner, Chicago, Ill., December 29, 1971.
[20] *Chicago Whip*, October 1, 1922.
[21] Dave Peyton, "Things in General," *Chicago Defender*, September 1, 1928.
[22] Peyton, "Things in General," *Defender*, May 22, 1926.
[23] Peyton, "Bad Habits," *Defender*, January 30, 1926.
[24] Nathan Irvin Huggins, *Harlem Renaissance* (New York: Oxford University Press, 1971), p. 9.

[25] Ross Russel, *Jazz Style in Kansas City and the Southwest* (Berkeley and Los Angeles: University of California Press, 1971), p. 9.
[26] Ibid., p. 3.
[27] Ibid., p. 10.
[28] Leonard, in *Jazz and the White Americans*, pp. 108–9, states that exterior economic pressures were primarily responsible for the commercialization of black jazz in the 1930s as opposed to internally motivated social and cultural aspirations on the part of black musicians.
[29] William E. Leuchtenburg, *The Perils of Prosperity, 1914–1932* (Chicago: University of Chicago Press, 1958), p. 152.
[30] Leroi Jones (Imamu Baraka), *Blues People*, p. 109.
[31] Ralph Ellison, *Shadow and Act* (New York: Random House, 1963), p. 249.
[32] Harlem stride pianist James P. Johnson and his student, friend, and disciple, Fats Waller, present an example of black jazz musicians who were genuinely attracted to European classical music for aesthetic reasons. Samuel Charters and Leonard Kunstadt noted, in *Jazz: A History of the New York Scene* (New York: Doubleday, 1962), p. 276, "They felt themselves to be in artistic agreement about the music they wanted to play. Not only were their styles very similar but both of them were deeply concerned with classical music and with composition." In an inerview conducted by Tom Davin Johnson stated: "From listening to classical piano records and concerts, from friends of Ernest Green such as Mme. Garret, who was a fine classical pianist, I would learn concert effects and build them into Blues and rags.... When playing a heavy stomp, I'd soften it right down–then, I'd make an abrupt change like I heard Beethoven do in a sonata.... Once I used Liszt's "Rigoletto Concert Paraphrase" as an introduction to a stomp.... I had gotten power and was building a serious orchestral piano. I did rag variations on the *William Tell Overture*, Grieg's *Peer Gynt Suite* and even a *Russian Rag* based on Rachmaninoff's *Prelude in C Sharp Minor*, which was just getting popular then." See "Conversation with James P. Johnson," in Martin Williams, ed., *Jazz Panorama*, pp. 52–3. Unfortunately, both Johnson's and Waller's careers ended in tragedy; neither of them was ever appreciated in his most serious and earnest mood, when inspired by the classical music that he loved.

References

Alexanderson, Gerald L. et al., 1990, *More Mathematical People Contemporary Conversations* Boston: Harcourt Brace Jovanovich

Blackett, P.M.S. 1949, *Fear, war, and the bomb: military and political consequences of atomic energy.* New York: Whittlesey House

Braham, Randolph L., 1981, *The Politics of Genocide*, New York, Columbia University Press

Chorin, Alexandre, 1994, *Vorticity and Turbulence*, Volume 103, Springer Science & Business Media

Colella, Phil, 2003, *The History of Numerical Analysis and Scientific Computing*, Society for Industrial and Applied Mathematics, Interviews with Peter Lax

Courant, Richard
    1947, *Differential and integral calculus*, New York, Interscience Publishers
    1953-62, *Methods of mathematical physics*, New York, Interscience Publishers
    1977 (with K.O. Friedrichs) *Supersonic flow and shock waves*, New York, Springer-Verlag
    1996 (with Herbert Robbins) *What is mathematics?: an elementary approach to ideas and methods.* New York : Oxford University Press, 2nd ed. / revised by Ian Stewart

Davis, Chandler, 1988, "The Purge," in *A Century of Mathematics in America, Part I*, Providence, Amer. Math. Soc., 413–428

Davis, Martin, 2013, personal communication.

Donaldson, Simon, 2011, "On the work of Louis Nirenberg", *Notices of the AMS* 58 (3) 469–472, March

Fermi, Laura, 1971, *Illustrious Immigrants*, Chicago, University of Chicago Press

Gelfand, I.M.
    1960, "On some problems of functional analysis", *Amer. Math. Soc. Transl.* (2), 16 (1960), 315–324.
    1963, "Some questions of analysis and differential equations", *Uspehi Mat. Nauk* 14 ... (2) 26, 201-219.

Gleason, Abbott, 2010, *A Liberal Education*, Cambridge, Tide Pool Press

Handler, Andrew, 1982, *The Holocaust in Hungary*, University, Alabama, The University of Alabama Press

Hargittai, Istvan. 2010, "The Last Boat from Lisbon: Conversations with Peter Lax", *The Mathematical Intelligencer*, 23:3, Fall, 24–30.

Hersh, Reuben
    1963 "Mixed problems in several variables," *J. Math. Mech.* 12 (3)
    1968 "Finite difference approximations for mixed initial-boundary value problems of general type," *SIAM J. on Numer. Anal.* 5 (2), June
    1979 (with T. Kato) "High-accuracy stable difference schemes for well-posed initial-value problems", *SIAM J. Numer. Anal.* 670–682
    1981 (with Philip J. Davis) *The Mathematical Experience*, Boston, Birkhauser
    1993 (with Vera John-Steiner) "A visit to Hungarian mathematics", *The Mathematical Intelligencer* 15 (2) 13–26
    2003, "The birth of random evolutions", *The Mathematical Intelligencer* 25 (1), 53-60
    2006, Review of "Selected Papers of Peter Lax", *Bulletin of the American Mathematical Society*, 43 (4) , 605–608, October
    2014, *Experiencing Mathematics What do we do, when we do mathematics?*, American Mathematical Society

Hoffman, Paul, 1998, *The Man Who Loved Only Numbers*, New York, Hyperion

REFERENCES

Holden, Helge, 2005, "Peter Lax Elements from his contributions to mathematics" http://www.abelprize.no/c53864/binfil/download.php?tid=53924

Hörmander, Lars. (with J. J. Duistermaat) 1972, "Fourier integral operators II". *Acta Math.* 128, 183–269.

Kazdan, Jerry et al., 2014, "Paul Roesel Garabedian 1927–2010", *Notices of the American Mathematical Society* 61 (3) 244–255. March

Kemeny, John, 1972, *Man and the computer*, New York, Scribner First Edition

Komoroczy, Geza, 1999, *Jewish Budapest*, Budapest, Central European University Press

König, Dénes, 2012, *Theory of finite and infinite graphs*, Boston, Birkhauser

Kreiss, Heinz, 1970, "Initial boundary value problems for hyperbolic systems", *Comm. Pure Appl. Math* 23, 277–298

Lanouette, William, 1992, *Genius in the shadows*, New York, C. Scribner's Sons

Lax, Anneli
    1965, "The New Mathematical Library A project to promote good elementary mathematical exposition", *The American Mathematical Monthly* 72 (9) 1014–1017 November
    1995, Cover photo, *The American Mathematical Monthly*, 102 (2) February

Lax, John, 1974, "Chicago's Black jazz musicians in the Twenties: Portrait of an era", *Journal of Jazz Studies* 1,2 113-118, June

Lax, Peter D.
    1944, "Proof of a conjecture of P. Erdos on the derivative of a polynomial", *Bull. Amer. Math. Soc.* 50, 509–513
    1952, "On the existence of Green's function," *Proc. Amer. Math. Soc.* 3, 526–531
    1956 (with R. D. Richtmyer) "Survey of the stability of linear finite-difference equations", *Comm. Pure Appl. Math.* 9, 267–293
    1957 "Hyperbolic systems of conservation laws II", *Comm. Pure Appl. Math.*, 10, 537–566
    "Asymptotic solutions of oscillatory initial value problems", *Duke Math. J.* 24, 627–646
    1962, (with R. S. Philips) "The wave equation in exterior domains", *Bull. Amer. Math. Soc.* 68 47–79
    (with C. S. Morawetz and R. S. Phillips) "Exponential decay of solutions of the wave equation in the exterior of a star-shaped obstacle," *Comm. Pure Appl. Math.* 16, 477-486
    1964 (with B. Wendroff), "Difference schemes for hyperbolic equations with high order of accuracy", *Comm. Pure Appl. Math.* 17, 381–398
    (with R. S. Phillips) *Scattering Theory*, Amer. Math. Soc. Bull. 70 130–142
    1966 (with L. Nirenberg) "On stability for difference schemes, a sharp form of Garding's inequality," *Comm. Pure Appl. Math.* 19, 473–492
    1967 (with J. Glimm), "Decay of solutions of systems of hyperbolic conservation laws", *Bull. AMS* 73, 105
    (with R. S. Phillips, *Scattering Theory*, New York, Academic Press
    1968, "Integrals of nonlinear equations of evolution and solitary waves", *Comm. Pure Appl. Anal.* 21 467–490
    1970 (with J. Glimm) "Decay of solutions of systems of nonlinear hyperbolic conservation laws", *Memoirs of the AMS* 101
    1972, "The formation and decay of shock waves," *Amer. Math. Monthly* 79, 227–241
    1973, Hyperbolic systems of conservation laws and the mathematical theory of shock waves, Conf. Board of the Mathematical Sciences, *Regional Conf. Series in Appl. Math.* (SIAM) 11
    "The differentiability of Polya's function," *Adv. Math.* 10, 456–465
    1976 (with R. S. Phillips), *Scattering theory for automorphic functions*, Ann. Math. Studies 87, Princeton University Press and Univ. of Tokyo Press
    (with S. Burstein and A. Lax) *Calculus with applications and computing*, Undergrad texts in math 1, Springer
    1977, "The bomb, Sputnik, computers and European mathematicians", *The Bicentennial Tribute to American Mathematics*, San Antonio, 1976, Math. Assoc. of America, 129–135
    1983 Report of the Panel on Large Scale Computing in Science and Engineering / Peter D. Lax, chairman. Washington, D.C.: National Science Foundation
    1986, "Mathematics and its applications," *The Mathematical Intelligencer* 8, 14–17
    1988, "The flowering of applied mathematics in America," *AMS Centennial Celebration Proc.*, 455–466; SIAM Rev. 31, 65–75
    1997, *Linear Algebra*, John Wiley & Sons, New York

1998, "The beginning of applied mathematics after the Second World War," *Quart. Appl. Math.*, 56, 607–615

2002, *Functional Analysis*, Wiley Interscience, New York

"Richard Courant," National Academy of Sciences *Biographical Memoirs* 82

2005, "John von Neumann: The early years, the years at Los Alamos and the road to computing", SIAM News, March 1

Selected Papers, Volumes I and II, New York, Springer

2006, *Hyperbolic partial differential equations*, American Mathematical Society and Courant Institute of Mathematical Sciences, Providence and New York

2012 (with Lawrence Zalcman) *Complex Proofs of real theorems*, American Mathematical Society, Providence

2014 (with Maria Terrell) *Calculus with applications*, New York, Springer

Levinson, Zipporah (Fagi), "From Revere to Cambridge," in *Recountings Conversations with MIT mathematicians*, Ed. Joel Segel, Wellesley, A. K. Peters

Li, Yan Yan, 2010, "Work of Louis Nirenberg," *Proceedings of the International Congress of Mathematicians*, Hyderabad, India

Macrae, Norman, 1999, *John von Neumann: The Scientific Genius Who Pioneered the Modern Computer, Game Theory, Nuclear Deterrence, and Much More*, Providence, American Mathematical Society

Marx, George, 2001, *The Voice of the Martians. Hungarian Scientists Who Shaped the 20th Century in the West*, Budapest, Akademiai Kiado

New York University *Alumni News*, 1970, "Cambodia, Student Deaths Cause Strong Reactions Withn NYU Community," "Class of '70 Hears Nobel Winner, Student at 'Convocation for Peace'," "Students, Faculty Express Views on National Crisis," XV (9), 1–5, June

New York University *News Bureau*, 1970, "The disruptions at Loeb, Courant, and Kimball", 1–24, September 23

Nirenberg, Louis, 2002, "Interview with Louis Nirenberg", *Notices of the AMS* 49 (4) 441-449, April

Niven, Ivan, 1995, "Yueh-Gin Gung and Dr. Charls Y. Hu Award for distinguished service to Anneli Lax", *The American Mathematical Monthly* 102 (2) 99-100, February

Oelsner, Lesley, 1970. "2 indicted in raid on N.Y.U. Center," *The New York Times*, July 30

Osher, Stanley "What sparsity and l1 optimization can do for you", people.ee.duke/edu/ icarin/SAHD_Osher.pdf

Patai, Raphael, 1996, *The Jews of Hungary*, Detroit, Wayne State University Press

Péter, Rózsa,

1964, *Playing with infinity*, New York, Atheneum

1967, *Recursive functions*, New York, Academic Press

1990, "Mathematics is beautiful," *The Mathematical Intelligencer* 12 (1) 58–64

Porter, Mason A. et al., 2009, "Fermi, Pasta, Ulam and the birth of experimental mathematics," *American Scientist* 97 (3) 214–221, May-June

Rademacher. Hans and Otto Toeplitz, 1957, *The enjoyment of mathematics*, Princeton University Press

Raussen, Martin and Christian Skau, 2006. "Interview with Peter D. Lax", *The European Mathematics Newsletter*, reprinted in the *Notices of the American Mathematical Society* 53 (2), February, pp. 223–229

Reid, Constance, 1976, *Courant in Gottingen and New York*. New York, Springer-Verlag

Richtmyer, Robert D. and K. W. Morton, 1967, *Difference methods for initial-value problems*, New York, Interscience Publishers

Sakamoto, Reiko, 1982, *Hyperbolic boundary value problems*, translated by Katsumi Miyahara Cambridge [Cambridgeshire]; New York: Cambridge University Press

Saul, Mark, 2000, "Anneli Cahn Lax (1922–1999)" *Notices of the AMS* 47 (7) 766–769, August

Smoller, Joel and Blake Temple, 2003, "Shock wave cosmology inside a black hole," *Proceedings National Academy of Sciences* 100 (20) 1216–1218, September30

Ulam, S. M., 1976, *Adventures of a mathematician*, Charles Scribner's Sons, New York

Von Kármán, Theodore, 1967, *The wind and beyond*, Boston, Little, Brown and Company

v. Neumann, John, 1958, *The computer and the brain*, Yale University Press, New Haven

Index

n-dimensional, 126
   inner product, 126
   phase space, 126
3-manifolds, 110
3-space, 126

a lovely old lady, 35
Abel
   Committee, 94
   Laureate, 94
   Niels Henrik, 41, 92
   Prize, 13, 92, 93, 114
Abel's life, 95
Aberdeen Proving Grounds, 169
Abramson, Jill, 42
abstract systems, 113
academies of science, 92
accuracy, 141
acoustics, 117, 118
ADD, 42
additive, 130
Adelphi University, 35, 161
Adirondack Mountains, 38, 45
Adriatic
   coastline, 1
   Sea, 2, 4
aerodynamics, 31, 32
aeronautical engineer, 119
aesthetically beautiful, 28
aggie, 20
air force, 52, 78
airfoils, 118, 119
airplane wings, 112
Airy, George, 143
Alamogordo Bombing and Gunnery Range, 28
Albania, 1
Albuquerque, 25, 30, 159
Alexandrov, Paul, 178
algebra, 110, 111, 126, 128, 133
   abstract, 32, 88
   linear, 105, 120, 177
   matrix, 113, 126
algebraic topology, 32

algebraically, 131
algebraist, 113
algorithm, 32, 118, 120, 133
   Metropolis, 121
   very fast, 107
Allgemeine Elektrizitatsgesellschaft, 11
Allies, 6
alternating method, 176
altitude, 130
aluminum, 131
Amazon Books, 109
Ambrose, Warren, 101
America, 12
American, 37, 92
   consul, 12
   history, 42
   politics, 42
American Bible Society, 33
American College of Surgeons, 31
American Legion, 42
American Mathematical Monthly, 81, 107, 163
American Mathematical Society, 37, 80, 82, 87, 91, 111, 120
American Philosophical Society, 121
American Revolution, 56
amplitude, 124
amusing, 117
analysis, 111
   complex, 110
   functional, 47, 53, 55, 66, 87, 114, 118, 128, 150, 153
   linear numerical, 105
   numerical, 63, 118
   vector, 108
analytic continuation, 112
angle, 126
   of attack, 112
angry, 116
angular velocity, 106
Annals of Mathematics, 109
anti-Semitic, 4, 102
   party, 5
anti-Semitism, xv, 5, 12

233

anti-snob, 41
antiballistic missile, 27
antikinks, 147
antisubmarine, 169
antisymmetric, 146
antiwar march, 78
Apfelstrudel, 102
applicability, 118
applications, 117, 121
applied mathematicians, 113
applied mathematics, xviii, 62, 113, 114, 116–118
Applied Mathematics Panel, 50, 180
appropriate norm, 66
approximate quadrature, 108
approximation, 66
    difference, 89, 111, 118
    finite difference, 89, 111, 117, 119, 133
    linear, 130
    methods, 131
    numerical, 124
Arany, Daniel, 8
arbitrarily high order, 89
arc, 125
area-filling curve, 154
Argonne Laboratory, 24
Armenians, 169
arms control, 30
army, 19, 52
Aronszajn, 85
ARPANET, 86
Arrow Cross, 5, 6, 14
art deco, 19
Art Nouveau, 114
artificial
    compressibility, 64
    viscosity, 135, 139, 140, 171
Artin's paradise, 106
Artin, Emil, 88, 98, 99, 106, 178
Artin, Natascha, 98
asbestos, 122
Aschbacher, Michael, 117
Asgeirsson, Leifur, 180
assembly language, 33
Astor Place, 33, 37
ASTP, 20
astronomy, 101
astrophysicist, 101
astrophysics, 32
asymptotic, 54
    analysis, 64
    behaviors, 84
    descriptions, 118
    expansion, 54
"Asymptotic solutions of oscillatory initial value problems", 54
Atiyah, Michael F., 22, 94, 113
Atlantic Ocean, 14

atmosphere, 30
atomic bomb, 20, 29, 30, 50, 57, 180
Atomic Energy Commission, 30, 51, 52, 61, 180, 181
atomic nuclei, 151
atomic spy, 25
attention span, 40
Auer, Leopold, 46
Augusta, 40
Auschwitz, 6
Ausgleich, 2
Australia, 8
Austria, 2
Austrians, 2
Austro-Hungarian Empire, 2, 4
Award for Distinguished Service, 163
awards, 41
axiomatic treatment of hyperbolicity, 111
axiomatization, 168
axis, 126

bacteria, 84
Baer, Reinhold, 180
Baker, Nicolas, 27
Balász, Béla, 114
balayage, 176
Baldwin, Alec, 19
Balkan, 2
ball game, 102
Banach, Stefan, 109
bankers, 113
Barney's, 19
Bartok, Béla, 6
Baruch, Bernard, 30
baseball, 97, 102
BASIC, 27
basic training, 19
basis vectors, 127
Battelle Institute, 61
beautiful, 35
    opera singers, 95
Beckenbach, Edwin F., 163
Beethoven, 182
Beijing, 92
Belgium, 5
Bell Labs, 150
Bellman, Richard, 22, 25, 163
Bergen, 94
Berger, Agnes, 11
Berkeley, xxi, 30, 64, 79
Berkowitz David, 47
Berkowitz Jerry, 46
Berkowitz Lori, 45–47
Berkowitz Susan, 47
Berlin, 6, 35, 88, 168, 169
Berlin, Irving, xix
Bernays, Paul, 88
Bernstein, Dorothy, 101

berry bushes, 164
Bers, Lipman, 99, 156
Bethe, Hans, 22, 170
Bible House, 33, 36, 37
Bieberbach conjecture, 119
Big Bang theory, 141
Bikini, 30
binary, 27
Birkhoff, 178
Birkhoff Prize, 91
Birkhoff, Garrett, 91, 156
Birkhoff, George David, 91, 102, 119
black hole, 141
black jazz musicians, 42
Black Panthers, 73–75
blacklist, 57
Bloch, Andre, 99
blood flow, 79
blows up, 131
blowup, 53, 131
blue Danube, 2
Bobisud, Larry, 61
Bobst University Library, 77
Bohr, Harald, 176
Bohr, Niels, 27, 28, 176, 178
Bolshevik, xvi
  Revolution, 4
Bolsheviks, 4, 75
Bolyai, Janos, 8
bomb, 169
  calculations, 31
  project, 137
bomb scare, xvi
bomb squad, 76
Bombieri, Enrico, 100
boo, 163
Boole, George, 113, 114
Born, Max, 88
Bosnia, 4
Boulder, 61
boundary conditions, 123
boundary-value problem, 123
Bourbaki, 114
Boussinesq equations, 148
Boussinesq, Joseph, 143
Braunschweig, 49
breather, 148
Breslau, 175
Brezis, Haim, 176
bridge, 66
Bronstein, Raphael, 46
Bronx, 40, 55
Brookline, Mass., 43
Brooklyn Polytechnic Institute, 79
Brouwer's Fixed Point Theorem, 116
Browder, Felix, 176
Brown University, 42, 85
Brown, Dr. Roscoe C., 75, 77

Brown, Jocelyn, 38
Bryson, Thomas, 44
buckets of rain, 94
buckling of a column, 20
Budapest, xv, xvi, 1–4, 6, 13–15, 31, 87,
    109, 114, 168
  Commodity and Stock Exchange, 4
  Jews, 4
  Sunday Circle, 114
  University, 8
Budapest-Pécs Railway, 4
Buffalo Thunder, 159
Bulgaria, 1
*Bulletin of the American Mathematical
    Society*, 109
Burstein, Sam, 82, 88
Burton, W., 111
Busemann, Herbert, 180
Bush, Vannevar, 49, 180
Buys, Mutiara, xix
Byzantine Empire, 1

Calculus, 163
calculus, 14, 83, 84
  differential, 122
  integral, 122
  of variations, 100, 110, 178
  reform, 84
  stochastic, 86
Calculus with Applications, 84
Calculus with Applications and Computing,
    82
Calderón, Alberto Pedro, 54, 111
California, 31, 52, 89
Caltech, 117
Calvinists, 1
Cambodia, 69–71, 77
Cambridge, 43
Camp David, 71
Canada, 45, 50, 120
canal, 142
cane, 39
Cantor's set theory, 8
capacity, 141
capitalistic system, 26
Cappell, Sylvain, 79
Carpathian Mountains, 2
Cartan, Elie, 109
Cartesian geometry, 126
Cartter, Alan M., 73, 75
CAT scan, 121
category theory, 32
Catholic, 27
  hierarchy, 5
Cauchy problem, 86, 179
Cauchy-Riemann, 110
Cayley, Arthur, 113
CDC 6600, 51, 61, 69, 72, 74, 75, 79, 181

CDC 7600, 79, 85
Census Bureau, 52
Center of Atmospheric Research, 85
Central Park, 19
Central Park West, 55
CERN, 151
CFL, 178
chain
    reaction, 9, 30
    rule, 138
Chakravarthy, Sukumar R., 82
Challenger, 119
champagne, 65
channel, 142
characteristic
    curves and surfaces, 54
    surfaces and rays, 111
Chauvenet Prize, 81, 85, 91, 135
Chauvenet, William, 81
chemical engineering, 168
chewing gum, 12
Chi, Professor Emile, 75, 76
Chicago, 30, 40, 42, 43
child prodigy, 119
Chinese takeout, 39
Chorin, Alexandre, xxi, 61, 62, 79, 121
Choroszczanski, 63
Christians in New York, 33
CIA spooks, 113
clashes, 116
classical
    physics, 117
    wave propagation, 118
classified project, 25
Clay Foundation, 32
cleaning, 45
Closed Graph Theorem, 160
closedness, 54
cnoidal waves, 143
Codazzi-Mainardi, 110
code breaking, 180
coefficients, 128
Cogburn, Bob, 60
cognitive dissonance, 128
Cohen, Paul, 59, 100
cold abstractions, 108
Cold War, 67, 181
Colella, Phil, 32, 86, 112, 116, 121
collaboration, 49
Collège de France, 94
College Station, Texas, 20
Colorado, 85
Columbia University, xvi, 17, 42
communism, xvi, 4
communist, 26
Communist International, 4
Communist party, 85
commutator, 142, 149

compact, 176
compactness, 139
competitive, 39
complete integrability, 143
completely integrable, 143, 146
    nonlinear, 148
    system, 148, 150
complex
    characteristic coordinates, 112
    domain, 112
    variables, 35
Complex Proofs of Real Theorems, 112
complexity, 116
compressible
    flow, 32, 137, 140, 141, 170
    fluids, 137, 141, 180
    media, 129
compression wave, 136, 137
computation, 121, 145
    practical, 89
computational, 89
    fluid dynamics, 61, 62, 170
    mathematicians, 116
    mathematics, 120
    science, 33, 120
computations, 58
computer, 31, 32, 51, 105, 118, 133, 145
    code, 118
    experiments, 142
    modeling, 124
    programming, 83, 84
    proof, 117
    science at NSF, 86
    science department, 80, 121
computing, 22, 25, 51, 83, 121, 129, 133
Computing Lab, 75
conformal mapping, 119, 176
conic sections, 105
Connes, Alain, 113
conservation
    form, 134, 141
    laws, 88, 137, 139–141
    of energy, 131
    of information, 123
conserved quantities, 143, 145, 146
consistency, 53
consistent, 53
constant, 131
    coefficients, 58, 134, 170
    speed, 124
contact discontinuities, 140
contacts, 140
contained fusion, 119
contempt of Congress, 56
contiguous points, 20
continuity, 54
continuum, 117
Continuum Hypothesis, 60

continuum mechanics, 131
Convair, 97
converge, 53
convergence, 53, 54, 84
convergent Fourier expansion, 128
convergent sequence, 83
converges, 89
Convocation for Peace, 77
cooking, 45
Cooper Square-4th Avenue, 33
Cooper Union, 33
coordinate system, 126, 127
Copenhagen, 27
copper, 122, 131
Cormack, Alan, 121
Cornell University, 84
corporal, 31
correct in the sense of Petrovsky, 58
cosine, 126–128, 131
cosmology, 141, 142
Council for a Livable World, 30
coupons, 14
Courant Institute, 20, 33, 41, 43, 46, 50, 57, 64, 72, 74–76, 79, 86–89, 94, 105, 111, 112, 120, 156, 157, 159
Courant, Ernst, 17, 46, 50
Courant, Gertrude, 17, 46
Courant, Hans, 17, 46
Courant, Leonore (Lori), 17
Courant, Lori, 46
Courant, Nina, 43, 46
Courant, Papa, 46
Courant, Richard, xvi–xviii, 17–19, 22, 33, 35, 37, 43, 46, 49–51, 53, 55, 57, 61, 86, 88, 97, 98, 105, 109, 135, 136, 160, 161, 176–179
Courant, Sara, 50
Courant-Friedrichs-Lewy, 51, 120
Courant-Hilbert, xvi, 55, 177–179
Cray, Seymour, 61, 79
crisis, 168
criticality, 24
Croatia, 1
Croatians, 1
crown prince, 3
Curtis, Kent, 86
curve in 4-space, 126
cylinder, 24
czar of Russia, 4
Czechoslovakia, 4, 5

Dafermos, Constantine, 68
Danbury Prison, 56
dark energy, 141
Dartmouth, 27
Dautray, Robert, 24
David, 45
Davis, Chandler, xvi, 55, 87

Davis, Martin, 74–76, 85
Davis, Phil, 85
Davis, Virginia, 74
de Vries, Gustav, 143
Debye, Peter, 177
Dee Dee, 39, 41, 45
degrees of freedom, 125
Denmark, 5
Department of Ordnance, 169
Depression, 150
depression, 179, 182
derivative, 153
Descartes, René, 128
determinants, 106
determinental representation, 111
detonations, 115
Devine, John, 162
diabetes, 47
diabolic, 58
Die Fledermaus, 2
difference
    approximations to PDE, 111
    equations, 53
    method, 141
    schemes, 32, 53, 54, 61
different style, 116
differential
    equation, 53
    operators, 54
diffraction, 33, 54, 55
diffuse, 123
diffusion, 122, 123
diffusive scheme, 81
diffusivity, 122, 131
dilemma, 129
dimension, 125
dining room, 19
direct problem, 115, 152
direct scattering computation, 146
direction, 126, 129
director, 79
Dirichlet problem, 108
Dirichlet's principle, 37, 175, 176
Dirichlet's Principle, Conformal Mapping, and Minimal Surfaces, 179
discontiniuity, 137
discontinuities, 32, 115, 152
discontinuous, 124
discretized Laplace equation, 86
discretizes, 32
dispersive method, 135
dissident, 56
distant stations, 115
divergence
    and curl of a vector field, 106
    of Fourier series, 108
Dodd, 147
domain decomposition, 176

Donsker, Monroe, 101
double crown, 2
Douglas, Jesse, 102, 179
Douglis, Avron, 180
drop of cream, 123
drunken driver, 43
Duistermaat, Hans, 54
Dunaway, Faye, 19
Dunford, Nelson, 163
dynamical systems, 46, 73, 125
dyslexia, 40
Dyson, Freeman, 61

$e = mc^2$, 21
Eötvös, Baron Lorand, 8
east side, 19
Ecole Polytechnique, 183
economics, 129
economy, 130
Edinburgh, Scotland, 142
Eichmann, SS Lieutenant Colonel Adolf, 6
eigenfunctions, 178
eigenvalues, 105, 106, 145, 146, 157, 177, 178
   of symmetric matrices, 107
Eighth Street Delicatessen, 36
eightieth birthday, 87
Einstein, Albert, 9–11, 17, 20, 28, 169
Einstein-Szilard refrigerator, 9
Eisenstein series, 153, 154
elastic collision, 147
Eldorado, 19, 38, 47, 87
Electric and Transportation Share Company, 4
electric chair, 12
electromagnetic waves, 153
electromagnetism, 54, 117, 118
electronic computer, 52, 122, 170
elegance, 116
elegant, 106
elementary
   calculus, 138
   particles, 54
Elizabeth, 42
ellipsoid, 24
elliptic, 110, 119
   curves, 113
Ellis Island, 14
embarrassing, 65
emperor
   Joseph II, 2
   of Austria, 1
   of Japan, 29
"Empty Bed Blues", 98
empty space, 141
enemy aliens, 14, 17
energy, 111, 140, 153
   crisis, 119
   inequalities, 111
engineering, 118, 129
England, 4, 179
English literature, 17, 55
Engquist, Björn, 61, 82
entropy, 139–141
   condition, 140, 141
Eötvös Competition, 8, 10, 13
Eötvös, Jozsef, 11
Eötvös, Lorand, 9
Epstein, Barbara, 85
Epstein, Jason, 85
equation
   Burgers', 139
   conservation, 122
   difference, 133, 170
   differential, 84, 108, 117, 137, 141
   diffusion, 123
   Einstein, 109
   elliptic, 66, 108, 178
   elliptic difference, 178
   elliptic partial differential, 34, 65
   Euler-Lagrange, 110
   finite difference, 120, 131, 140
   Hamilton-Jacobi partial differential, 110
   heat, 123
   hyperbolic, 178
   hyperbolic partial differential, 82, 108, 156
   integral, 137, 176
   inviscid Burgers', 133
   Kadomtsev-Lukashvili, 147
   Kadomtsev-Petviashvili, 148
   Korteweg-de Vries, 129, 142, 146
   Korteweg-de Vries partial differential, 34
   Lax, 149
   linear, 105, 130
   linear hyperbolic, 141
   linear partial differential, 126
   nonlinear, 128, 131, 134, 141, 142, 145, 147
   nonlinear differential, 129, 130
   nonlinear hyperbolic partial differential, 122
   nonlinear parabolic, 110
   nonlinear partial differential, 140, 170
   nonlinear Schrödinger, 147, 148
   one-dimensional heat, 122
   one-dimensional wave, 124
   parabolic, 123, 178
   partial difference, 118, 178
   partial differential, 25, 32, 34, 54, 64, 66, 87, 95, 108, 109, 111, 115, 118, 119, 121, 122, 131, 136, 137, 141, 146
   partial differential of hyperbolic type, 117
   quintic, 92
   Schrödinger, 109, 145
   simplified, 118

sine-Gordon, 147
soliton, 129
transport, 124, 131
wave, 131, 153
well-posed linear evolution, 89
equilibrium, 34, 122, 123
equivalence principle, 53
Erdelyi, Arthur, 8
Erdős, Paul, 8, 13, 17, 22, 24, 101, 116
ergodic hypothesis, 119
Erlangen, 88
Esquire magazine, 55
essentially non-oscillatory, 82
Esterházys, 1
estimate, 66
estimates, 66
Ethical Culture Society, 40
Euclidean
  geometry, 35
  space, 128
Euler, Leonhard, xv, 20, 32, 117, 122, 137
Europe, 44
European
  horrors, 108
  Jews, 169
Evangelical Gymnasium, 168
every mathematician's bookshelf, 108
exact solutions, 131
existence, 117
experimental outcome, 152
experimentation, 118
explicit
  formula, 117, 140
  solutions, 118
explosions, 169
extinguisher, 76
extortion, 74
extraordinary talent, 14
extraterrestrials, 24

Faculty of Theology, 8
Faddeev, Boris, 153, 154
falling in love, 87
Farkashazi, Dr. Victor, 11, 15
Farkashazi, Steven, 15
farmer, 101
Farmer, Henry, 27
Farr, Bill, 15
fascism, 5
Fast Fourier transform, 114
fast matrix multiplication, 120
federal dollars, 30
Federal Institute of Technology, 179
feedback, 129
Feferman, Sol, 59
Fejér, 3
Fejér, Leopold (Lipot), 8, 99
Father Ignatius Fejér, 8

Fejér, Ignatius, 8
Fejér, Leopold (Lipot), 8
Fekete, Michael, 8, 168
Feller, Willi, 180
fencers, 5
Ferdinand, Franz, 98
Fermi, Enrico, 22, 24, 27, 30, 100, 144
Fermi-Pasta-Ulam
  (FPU), 34, 145
  recurrence, 145
Feynman, Richard, 22, 170
fiber optics, 148
fictitious geometry, 125
Fields Medal, 92, 93
Fields, John Charles, 92
Fieldston School, 40
figure skater, 39
finance, 129
finite
  difference approximations, 88
  differences, 32, 131
  speed of propagation, 64
  time, 130
Finite Mathematics, 27
Finite-Dimensional Vector Spaces, 105
finite-mass cutoff, 141
fireball, 28
First World War, 4, 19, 169, 176, 177, 179
first-order
  methods, 141
  term, 54
fission, 118
Fiume, 4
fixed bayonets, 78
Flanders, Donald, 179, 182
Flaschka's variables, 148
flesh and blood, 108
Flexner, Abraham, 179
Florence, 87
Florida, 20
flowering, 143
  of applied mathematics, 82
flowings, 122
fluid, 122
fluid dynamics, xvii, xxi, 109, 112, 117, 118, 135, 140, 141, 170
flux of heat, 122
fluxions, 122
fools, 109
Ford Foundation, 181
Ford Prize, 91
Ford, Lester R., 65
formalists, 168
foundations, xvii
founding fathers, 108, 109
Four Color Theorem, 117
four-dimensional phase space, 125
Fourier

analysis, 134
analyst, 97
expansion, 128
formulas, 128
integral operator, 54
integral operator calculus, 55
integrals, 127
series, 127
transform, 54
Fourier, Joseph, 127
France, 4, 5, 12, 44, 109
fraulein, 8
Fredholm theory of integral equations, 178
Fredholm, Eric, 176
Freedom of Information Act, 85
freezing the coefficients, 134
French probabilists, 113
frequency, 54
domain, 54
Friedman, Bernard, 50, 180
Friedrichs, Kurt Otto, 33, 34, 37, 49–51, 54, 64, 100, 101, 109, 120, 133, 135, 136, 153, 178–180, 182
Frobenius, Ferdinand Georg, 88
Fuchs, Klaus, 30
Fuchs, Laszlo, 13
Fuhrer, 109
Fulbright Lecturer, 60
function
exponential, 84, 130
periodic, 128
space, 113, 126
Function Theory, 177
Functional Analysis, 14
functional analysis, 108, 111, 122, 126
Functional Analysis and Semigroups, 61
functions
automorphic, 88
continuous, 108
elliptic, 93
fur coat, 43
fusion, 120
bombs, 118

gadget, 28
Gallai, Tibor, 13
Galois, 92
Garabedian, Paul, 57, 85, 112, 119
garbage, 55
Garbo, Greta, 6
Gårding's inequality, 65
Gårding, Lars, 61
Gardner, Cliff, 91
Gardner, Clifford, 145
Gardner, Martin, 85
gastroenterostomy, 31
Gauss, Carl Friedrich, xv, 114
Gelfand, Israel Moiseyevich, xv, 99, 153

Gelfand-Levitan integral equation, 145
general hyperbolic systems, 152
General Motors, 15, 20
general relativity, 118, 141
geometric, 128
geometrical optics, 54, 55
geometrically, 125
geometry, 49, 142, 147, 179
differential, 110, 179
non-Euclidean, 109
projective, 99
George David Birkhoff Prize, 80, 120
Gerber, Porter Dean, 159
German, 109, 181
German-Italian axis, 5
Germans, 1, 14
Germany, 5, 11, 60, 100, 169, 179, 181
ghettos, 2
Gibbs, Josiah Willard, 114
Gleason, Abbott, 42, 44, 160
Glimm, James, 67, 140
gloom, 182
glucose, 47
Gödel, Kurt, 169
going mad, 109
Goldstein, Max, 79
golf, 27
Goodhue, William, 64
Gorky Park, 181
gossip, 3
Göttingen, xvi, 17, 19, 88, 98, 114, 168, 175, 177–179
Grad, Harold, 33, 50, 180
graduate curriculum, 108
grandparents, 40
graph, 130
gravitation, 128
gravitational, 126
gravity, 125
Great Depression, 4, 12
Greece, 1
Green's, 116
Greene, John M., 145
Greenglass, David, 25
Greenleaf, Frederick, 75
Greenwood, Dr. Marnie, 39, 42, 43, 45, 94
Greenwood, Priscilla, 60
Griego, Richard, 60
Gromov, Mikhail, 94
Groves, General Leslie, 53
Grundlehren, 177
guesswork, 114
guitar string, 124
Gymnasium, 11, 97

H-bomb, 27, 30
Haar, Alfred, 175
Habsburg, 2

emperor, 2
Empire, 4
Hadamard, Jacques, 112, 179
Hahn-Banach, 116
haiku, 150
haikus, 150
Hajós, 12
Halmos, Paul, 105, 116
Hamburg, 169
Hamilton, 114
Hamilton, Ontario, 65
Hamilton–Jacobi equations, 61
Hamilton-Perelman proof, 110
Hamiltonian mechanics, 143
Hanford, 21
Hardy, G. H., 102, 103, 178
Harish-Chandra, 100
Harlem Studio Museum, 38
Harten, Amiram, 81
Hartford, Connecticut, 43
Harvard, 77, 102
Hasse, Helmut, 98
Hasse-Mendelsohn, Herr, 98
Hauptman, Herbert, 121
Hausner, Mel, 74, 75
Hautermans, Mrs., 27
Haydn, Joseph, 1
heart attack, 47
heart models, 62
heat, 122
    conduction, 122
    energy, 122, 123
heavy water, 27
Hebrew, 65
Hebrew University, 168
Hedrick Lectures, 79
Heisenberg, xvii
Heisenberg, Werner, 28, 88, 99, 100, 151, 169
Heller, Alex, 22
Hellinger, Ernst, 175
Helton, J. William, 157
Herbert Hoover Presidential Library, 43
Herglotz, Gustav, 88
Hermansen, Tormod, 94
Hermiston, 142
Hersh, Reuben, 39
Hester, President James M., 69, 75, 77
high-dimensional space, 128
high-prestige, 113
high-speed
    computer, 107, 119
    computing, 113, 145
    computing devices, 170
higher-dimensional geometry, 126
Hilbert Space, xvii
Hilbert space, 108, 128, 149, 153, 156
Hilbert's tenth problem, 85

Hilbert, David, xv, xviii, 86, 88, 105, 114, 154, 168, 175
hillbillies, 20
Hille, Einar, 61, 150
Hindle Committee, 77, 78
Hindle, Brooke, 77
Hiroshima, 29, 30
Hitler's Germany, 5, 12
Hitler, Adolf, 5, 15, 31, 98, 109, 169, 179
Hitler-Mussolini-Tojo, 14
Hoffman, Paul, 3
Holden, Helge, 95
Holland, 5
Hollywood, 12
Holocaust, 1, 6, 13
Holy Roman Empire, 1
homogeneous, 54, 130
honorary Aryans, 98
honorary societies, 92
Hoover Institute, 101
Hopf, Eberhard, 139, 178
Hopf, Heinz, 179
Hopf, Heinz-Otto, 49, 140, 175
Hörmander, Lars, 54, 55, 59
Hornig, Donald, 28
horseback, 142
Horthy's Hungary, 5, 12
Horthy, Miklós, 4–6
Hotel Continental, 114
Hoyt, Frank, 28
huge white stretch limousine, 95
Hugoniot, Pierre-Henri, 140
human body, 115
Hungarian, 1, 37, 92, 116
    accent, 24
    General Credit Bank, 3
    Jews, 6
    miracle, 8, 87
    Nazis, 15
    River Navigation and Maritime Share Company, 4
    school, 108
    translations, 12
Hungarians, 2, 24
Hungary, 2, 4–6, 14, 15, 116
Hurwitz-Courant, 177, 182
Husserl, Edmund, 176
Huygens principle, 153
hydrogen bomb, 30
hydromagnetic flows, 33
hyperbolic, 58, 89, 110, 119, 123
    conservation laws, 111, 112, 139–141
    PDE, 64, 65
    problems, 64
    secant, 143, 148
    system, 58
Hyperbolic Partial Differential Equations, 111, 150, 151

Hyperbolic Systems of Conservation Laws, 81
hyperbolicity, 110

IBM, 27
   Yorktown Heights, 159
ICBM, 97
ice water, 122
image
   reconstruction, 61
Immersed Boundary Method, 79
impenetrable contributions, 112
implosion, xvii, 117, 169
   wave, 136
incompressible flow, 32, 137
Infante, Jim, 86
Infantry Replacement Training Center, 20
infinite, 130, 131
   decimals, 83
   regress, 129
   series, 128
infinite-dimensional, 128
   space, 128
infinitesimal generator, 153
infinity, 131
informatics, 80
initial
   boundary value problem, 61, 112, 123
   concentrated impulse, 123
   value problems, 141
   velocity, 124
Initial-Value Problems for Partial Differential Equations, 53
inner product, 126
insane asylum, 99
insanities, 109
instability, 131
Institute for Advanced Study, 17, 100, 169
Institute of Afro-American Affairs, 75
Institute of Theology, 102
insulin, 8
integrable
   nonlinear PDE's, 118
   systems, 111
integral expressions, 122
intensity, 124
intercontinental missiles, 119
interference, 54
internal memory, 27
International Congress of Mathematicians, 86, 92
   in Kyoto, 82
International Educational Board, 178
Internet, 86
intuitionists, 168
invasion of Normandy, 29
inverse
   problem, 115, 152
   scattering, 145, 146, 149
   scattering transform, 145
inviscid gas dynamics, 112
Iowa, 43
irregularity, 117
Isaacson, Eugene, 50
isobar, 138
Italy, 65
iterative methods, 107
Ito, Kyoshi, 86
Iwo Jima, 29

Jacobi, 149
Japan, 14, 24, 28, 66, 101
   surrendered, 29
Japanese
   civilians, 28
   mainland, 29
jazz singer, 98
Jemez Mountains, 21, 159
Jerusalem, 168
Jewish, xvi, 4, 49, 108, 168, 179
   Cultural Association, 13
   Gymnasium, 13
   Hospital, 6, 12, 87
   question, 4
Jews, 1, 2, 5, 37, 98, 102
John, Fritz, 33
John-Steiner, Vera, 87
JOHNIAC, 32
Joseph, Emperor Franz, 4
Josiah Willard Gibbs Lectureship, 80
Judson Hall, 50
jump discontinuity, 140
junger Amerikaner, 49

König, Dénes, 8
Kozepiskolai Matemaikai Lapok, 8
Kármán, Mor, 11
Kac, Mark, 64, 91
Kadar, 15
Kadomtsev-Petviashvili, 143
Kafkaesque, 85
Kakutani, Shizuo, 101
Karle, Jerome, 121
Kato, Tosio, 89
Katowice, 35
Kayton, Bruce, 75
KdV, 143
Keillor, Garrison, 19
Keller, Herb, 50
Keller, Joe, 33, 50, 54, 64, 79, 97, 159, 180
Kelvin, 114
Kemeny, John, 22, 26, 27
Kent State University, 69, 70, 78, 92
Kent, Ohio, 70
Keyfitz, Barbara, 64
Khinchin, Aleksandr, 178
Kiev, 98

Kimble Hall, 72
kindness, 64
kinematics, 125
    and dynamics, 106
kinetic energy, 153
King Oscar II, 93
kinks, 147
Kistiakowski, George, 22
Klein, Felix, 88, 105, 114, 175, 178
Klein-Gordon, 147
Kline, Morris, 54
Klipple, Edmund Chester, 20
knee replacement, 39, 48
knees, 48
Knopp, Konrad, 35
Kohn, Joe, 54, 66, 79, 80, 111
Kohn, Walter, 121
Kolmogorov, Andrey, 178
Kolmogorov-Arnold-Moser, 150
Kon, Mark A., 107
König, Dénes, 8, 13
König, Julius, 8
Königsberg, 168
Korda, Sir Alexander, 6
Korn, David, 112
Kornfeld, Aharon, 3, 9
Kornfelds, 3
Korodi, Albert, 9, 10
Korteweg, Diederik Johannes, 143
Korteweg-de Vries, xvii, 143, 145, 147
    equation, 86
Kotlow, Dan, 76
Kracow, 97
Krahn, Edgar, 180
Krantz, Steven G., 109
Kranzer, Herbert, 180
Krein, Mark, 98
Kreiss, Heinz-Otto, 64, 112, 116
Kremlinology, 4
Kronecker, Leopold, 8, 88
Kruskal, Martin, 33, 50, 142, 143, 145, 146, 180
Kummer, Ernst, 88, 98
Kun, Béla, 4
Kun, Miklós, 4
Kürschák, József, 168

La Jolla, 51
labor
    battalion, 15
    service, 6
laboratory, 118
Ladyzhenskaya, Olga, 67
LaGuardia Airport, 19
Lake Placid, 47
Lamy, 20
Lanczos, Cornelius, 8
Landauer, Rolf, 17

Laplace operator, 177
Laplace-Beltrami, 154
laughter, 116, 117
Lawrence, Kansas, 85
laws of nature, 114
Lax
    equation, 146
    equivalence principle, 53
    family, 13, 14
    Pair, xx, 142, 145–149
    Panel, 85
Lax, Anneli Cahn, 19, 33, 35, 38, 39, 44–47, 50, 65, 66, 76, 82, 84, 85, 88, 94, 160, 180
Lax, Henrik (Henry), 6, 12, 14, 15, 18, 19, 35, 38, 43, 44, 87
Lax, James (Jimmy), 18, 19, 37–39, 42, 44, 45, 47, 94, 164
Lax, John, 8, 18
Lax, John (Johnny), 18, 37, 42–45, 160
Lax, Klara, 6, 12, 15, 18, 38
Lax, Peter, 35, 38, 50
Lax, Pierre, 38, 94
Lax, Timmy, xix, 38, 94
Lax, Tommy, xix, 39, 94
Lax-Friedrichs, 133
Lax-Milgram, 156
    lemma, 133
Lax-Richtmyer, 53
    Equivalence Theorem, 160
Lax-Wendroff, 133, 134
    Method, 60
    Theorem, 60, 133
least square, 105
Lefschetz, Solomon, 181
LeMay, General, 29
Lenin, 4
Leray, Jean, 61
Leray-Schauder degree, 108
Leroy P. Steele Prize, 87
Lester R. Ford Award, 81, 135
Lesznai, Anna, 114
Leuchtenburg, William, 44, 210
level-set, 61
Levi, Beppo, 176
Levi, E. E., 161
Levinson, Norman, 102
Levinson, Zipporah (Fagi), 102
Lewis, A. S., 157
Lewis, Sinclair, 19
Lewy, Hans, 178, 180
lice, 109
Lie, Sophus, 93
Life magazine, 58
light, 118, 122, 152
    propagation, 128
    rays, 152
    waves, 117

Lilacs, 94
limit, 131
limiting value, 122
Lincoln Center, 19
linear, 131
  operators, 60, 149
  space, 108
Linear Algebra, 105
Linear Operators, 163
linearize, 145
Lisbon, 13, 14
little birdie, 22
Livermore, 30, 85
  Lab, 52
living room, 19
Loeb Student Center, 71, 72, 74
logarithm, 84
logic, 129
logicians, 117
London, 92
London, Jack, 12
long, straight cup of coffee, 123
Loon Lake, 38, 43, 45, 47, 65, 163, 164
Lorch, Lee, 87
Lord, 116
Lori, 43
Los Alamos, xv–xvii, 11, 21, 24, 26, 27, 30–33, 53, 85, 97, 114, 133, 136, 144, 159, 167, 169, 170
Lower East Side, 47
lox, 6
loyalty oath, 57, 84
Ludwig, Donald, 180
Lukacs, Georg, 114
lunch, 145
Lüneburg, Rudolf, 50, 180
Lutherans, 1
lyceum, 35
Lyusternik, Lazar, 178

Mac Lane, Saunders, 178
MacArthur Fellowship, 80
MacArthur, General Douglas, 52
machine computation, 114
Macrae, Norman, 3
Madison Square Garden, 77
magnetohydrodynamics, 32
magnitude, 126
Magyar nobility, 5
Magyars, 1, 2, 4, 5
Majda, Andy, 111
Manakov, 148
Manhattan, xvi, 17
  Project, xv–xvii, 11, 17, 20, 24, 28–30, 53, 133, 136
  School of Music, 46
MANIAC, xvii, 32, 144
Manin, Yuri I., 143

Marchant, 24
Marchisotto, Elena, 164
Marjorie Morningstar, 19
Martha's Vineyard, 39
Martians, 17, 20, 31
Marx, George, 3, 26
Marxist, 114
  Students for a Democratic Society (SDS), 75
Maschke, Heinrich, 175
masochism, 41
Math Monthly, 81
Mathematica, 116
mathematical
  algorithms, 119
  descendents, 89
  exactitude, 117
  physics, 117
Mathematical Association of America, 65, 79, 81, 91, 161
Mathematical Intelligencer, 114
Mathematical Society, 12
Mathematicians' Action Group, 87
mathematics
  applied, 24, 57, 86, 87, 119, 127, 180
  numerical, 105, 121
Mathematics Genealogy Project, 34, 160
Mathematics Institute, 19
Mathematischen Annalen, 177
Mathematischen Zeitschrift, 177
Matrices, 113
matrix, 146
  inequalities, 106
  operations, 107
Maugham, Somerset, 13
maximum principle, 66
Maxwell, 32, 114
Maxwell's equations, 54
Maxwell, James Clerk, 122
McCarthy, 84
McShane, E. J., 178
mechanical systems, 125
mechanics, 49, 126
  classical, 126
  continuum, 32, 109, 170
  fluid, 63
  Newtonian, 110
  quantum, 54, 88, 109, 118, 169
medical
  imaging, 115, 152
  school, 44
memorial service, 44
Mendelsohn, Felix, 98
Menshevik, 75
metal, 131
  bar, 122
meteorology, 31, 109
Methods of Mathematical Physics, 161, 181

Metropolis, Nicholas C., 32
Metropolitan Opera, 19
Michelangelo, 58
Michigan, 64
micro-local, 54
microlocal analysis, 54
Milgram, Arthur, 156
military, 135
Milnor, John, 55
minimal surfaces, 100
minister of culture, 11
Ministry of Education, Research and
    Church Affairs, 94
Minkowski, Hermann, 12, 175
Minta, 11, 27
misleading cliché, 113
MIT, 94, 101, 102
Miura, Robert M., 145
mixed initial boundary value problem, 152
model, 84
    example, 118
Model School, 11
modern mathematics, 113
Mollenauer, Linn, 149
Molnar, Ferenc, 6, 87
Molotov cocktails, 75
Monet, 7
monomaniac, 100
monopoly, 30
Monte Carlo, 121
    method, 27
Montreal, 34, 50, 65, 170
Moore, R. L., 20, 99
Mor, 3
Morawetz, Cathleen, 33, 36, 50, 61, 99, 136,
    159, 180
Morawetz, Herbert, 99
Morgenstern, Oskar, xvii
Morgenthau, Henry, 12
Morikawa, George, 74
Morton, K. W., 53
Moscow, 92
    University, 99
Moser, Gertrude, 43, 46
Moser, Jürgen, 43, 46, 73, 74, 76
Mosers, 43
mother's brother, 15
Mount Holyoke College, 42, 43, 164
movie, 145
moving fronts, 61
Mozart, Wolfgang Amadeus, 1
multi-resolution, 81
multi-soliton, 142, 145, 147
    solutions, 145
Munch, 95
Münster, 177
murdered, 14
Murray, Francis, xvi, 17, 169

Muttersprache, 97
my 1962 thesis, 112
my favorite, 108
My Lai, 69
my spies, 22
Mysticism and Logic, 113

Nachlass, 114
Nagasaki, 29
naked machines, 33
nanny, 8, 38
Napoleon, 2
narcolepsy, xx, 66
Nassau Street, 33
National Academy of Sciences, xvii, 17, 80,
    86, 89, 100
National Guard, 70, 78
National Medal of Science, 86, 182
National Research Council, 50
National Science Board, 85
National Science Foundation, 85, 180, 181
National Science Medal, 37
native tongue, 87
natural frequencies, 107
Nature, 102
nature, 183
Navier-Stokes, 110, 117
Nazi, 108, 109
    atomic bomb, 11
    Germany, 14, 28
Nazi fascism, xv
Nazis, 4, 5, 31, 35, 49, 98, 179, 181
Nem, nem, soha, 4
Neugebauer, Otto, 97, 180, 182
Neumann, 27
Neumann, Carl, 176
Neumann, Nelly, 176
neutron transport, 24
neutrons, 169
New Mathematical Library (NML), 163
New Mexican, 133
New Mexico, 30, 60, 78, 87, 159
New Mexico National Guard, 71
New Rochelle, 17, 46
New Year's Eve, 43
New York, xv, xvi, 2, 6, 15, 35, 40, 45, 49,
    64, 78, 92, 109, 114, 157, 159, 179
New York City, 14, 180
New York Review of Books, 85
New York University, xvi, 17, 18, 35, 70,
    71, 88, 161, 179
New Yorkers, 45
Newman, Donald, 112
Newton, Sir Isaac, xv, 54, 117, 122, 128
Niels Henrik Abel Memorial Fund, 94
Nielsen, Jakob, 178
ninety-seven years old, 41

Nirenberg, Louis, xix, 33, 34, 50, 54, 55, 65, 80, 111, 120, 159
Nixon, President Richard, 69, 71
no curfew, 101
Nobel Prize, 77, 93, 94, 121
Nobel, Alfred, 92
Noether, Emmy, 88, 98, 106
noise, 145
non-applied, 113
non-Euclidean geometry, 8
non-linear conservation laws, 86
noncombatants, 29
nonelephant zoology, 129
nonlinear, 68, 124, 131, 137, 138, 143
    conservation laws, 118, 129, 140, 141
    equations, 146, 170
    mathematics, 130
    PDE's, 131
    Schrödinger, 143
    springs, 144, 145
    system, 141
    waves, 142
nonlinearity, 113, 129, 131, 140
nonstandard analysis, 85
Norbert Wiener Prize, 62, 82
normalized, 126
North Vietnam, 70
Northridge, 164
Norway, 5, 92, 93, 115
Norwegian
    Academy of Science, 93
    Academy of Science and Letters, 94
    kroner, 94
    University of Science and Technology, 95
not all that difficult, 108
Notebooks of the Mind, 87
Novosibirsk, 65, 181
nuclear, 30, 169
    bomb, 21
    catastrophe, 30
    disarmament, 30
    explosion, 136
    fission, 11, 24, 29
    fusion, 24, 34
    reactor, 21
    weapons, 170
number theorist, 98
numerical
    analysis, 53, 60, 114
    analysts, 119
    approximation, 136
    approximation scheme, 53
    calculations, 61
    examples, 83
    experience, 68
    experimentation, 150
    experiments, 32
    mathematicians, 116
    methods, 64, 134
    model, 145
    result, 114
    schemes, 31
"Numerical solutions of partial differential equations", 65
numerus clausus, 13
NYU, xvii, 19, 20, 33, 36, 49–53, 60, 69, 70, 133, 180
    computing center, 79
    math department, 52
    News Bureau, 76
    Senate, 76, 77
    Student Union Building, 78

Oak Ridge, 19, 20
Oberwolfach, 114
oceanography, 110
office of Naval Research, 180
oil
    exploration, 115, 152
    reservoir, 115
Okinawa, 29
Oklahoma, 85
Oleinik, Olga, 66, 67, 133
Olum, Paul, 22
one big family, 67
operator
    singular integral, 111
    theory, 55
operators
    Calderón-Zygmund singular integral, 66
    pseudo-differential, 66, 111
    self-adjoint, 169
Oppenheimer, Robert, 22, 28, 30, 170
optical fibers, 150
orthogonal, 128
    coordinates, 128
    function systems, 177
    transformation, 146
oscillation, 145
oscillatory, 135
Osher, Stanley, 61, 82, 121
Oslo, xv, 13, 94, 114
OSRD, 49, 50, 180
Osserman, Bob, 59
Ottoman
    Empire, 1
    Turkey, 4
oven, 122
over-relaxation, 120
oversimplifications, 131

Péter, Rózsa, 8, 11–13
Painlevé, Paul, 112
painting, 114
Palo Alto, 101, 159
pancreas, 164
pancreatic cancer, 47

Papanicolaou, George, 60, 85
parabola, 125
parabolic, 58, 89, 110, 123
paradise, 106
paradox, 129
parameter, 125
Paris, 35, 92
   Academy of Sciences, 86
Parter, Seymour, xx
partial derivatives, 113, 122, 124
particles, 145, 151
partitioned, 145
past master, 101
patterns, 114
Pavlov, Ivan, 153, 154
peacenik, xv
Pearl Harbor, 14, 49, 101
peasant, 114
Peierls, Rudolf, 98, 170
Pencak, Bill, 42
Pennsylvania University, 92
periodic, 127, 128
Peskin, Charles, 62, 79
Petenyi, Dr. Géza, 15
Peter, 3
Petrovsky, I. G., 99
Petrowsky-correct, 110
Ph.D. students, 89
phase space, 113, 125
Phillips, Ralph, 59, 61, 66, 67, 82, 88, 109, 150, 151, 153
physical intuition, 118
physicists, 121
physics, 55, 110, 113, 117, 118, 121, 131, 142, 147, 151
   modern, 88
Picasso portrait, 26
pickle barrel, 97
Ping-Pong, 39
Pinsky, Mark, 60
piston, 136
Placzek, 27
plasma physics, 32
plasmas, 33
Plateau's problem, 179
Platonic ideals, 83
Plattsburg, 45
Playing with Infinity, 11
plucking, 124
plutonium, xvii, 21, 28, 136, 169
   bomb, 117
pocket calculators, 83
Poincaré, Henri, xv, 88, 114, 176
Poincaré-Thurston, 110
Pojoaque, 159
Polányi, Michael, 11
Poland, 5, 35, 55
Poles, 109

Dr. Ferenc Polgár in Geneva, 87
Pólgar, Ferenc, 12
Polish, 108, 109
Politzer, 3
Pólya function, 154
Pólya, George, 8, 31, 133, 154, 168
Pólya, Jenö Sándor, 31
Pope John Paul II, 176
Pople, John, 121
positive schemes, 133
potential energy, 153
Prandtl, Ludwig, 175
pre-FORTRAN, 33
presidency, 27
Pressman, Barney, 19
pressure, 122, 135
   wave, 136
Price, Ray, 71
Prime Number Theorem, 112
prince, 95
Prince Camp, 45
Prince of Computing, 114
Princeton, 17, 26, 80, 100, 102, 157, 169, 181
   University Press, 82, 109
principle of uniform boundedness, 53
prizes, 41, 91, 119
probabilistic, 178
probabilists, 60
problem solving, 13
Procrustean bed, 120
programmed, 33
progressing waves, 64, 111
proofs, 116
propagation of singularities, 64, 111
Prussia, 2
Prussian Germany, 4
Prussian Ministry of Education, 178
pseudo-differential, 54, 64
pseudosphere, 147
psycholinguist, 87
pulse of heat, 123
pure, 113, 116
   mathematicians, 116
   mathematics, 113
purity, 88
Putnam, Professor, 35

quadric surfaces, 105
quantum
   field, 150
   mechanical, 151
   mechanics, 152, 153
   theory, 100
queen, 95
quotas, 5
quotient spaces, 106

Rácz, László, 168

rabbi of Lemberg, 97
racetrack, 45
radar, 180
Rademacher, Hans, 11
radioactive elements, 84
radiology, 12, 45
Radó, Tibor, 179
Radon-Nikodym theorem, 108
Raman, 148
Rand Corporation, 25
Rand, James, 52
random choice, 67
random evolution, 60, 85
Rankine, William, 140
Rankine-Hugoniot, 137
rarefaction
    rate, 140
    wave, 137
rate of convergence, 107
rates of change, 122
Rauch, Jeffrey, 64
Raussen, Martin, 114
ray, 139
Rayleigh, 114
Reagan, Ronald, 28
real
    computer, 117
    numbers, 83
    plane curves, 111
    world, 119
rectangular hyperbola, 130
recursive functions, 11
recursively, 118
Red Square, 97
reentry problem, 119
reform of calculus teaching, 82
Regent, 4
Regents of the University
    of California, 84
regularity, 117
Reichel-Pólya, 31
Reid, Constance, 183
relativistic mechanics, 34
relativity, 109
religion, 6
Rellich, Franz, 180
Remington Rand, 52
renormalization, 68
representations, 117
resolution, 141
resolving power, 141
Return of the Native, 17
Ricci flows, 110
Richardson, James, 32
Richman, Chaim, 31
Richtmyer, Bob, 53, 139
Riemann
    hypothesis, 154
    matrices, 143
    problem, 68, 140
    surfaces, 143, 177
Riemann, Bernhard, xv, 114, 137, 139, 140,
    175, 176
Riesz representation theorem, 156
Riesz, Frederic, 176
Riesz, Frigyes, 8, 108
Riesz, Marcel, 8
righteous gentiles, 15
rigid
    body, 106
    rods, 126
rigor, 88, 114, 117, 118
rigorous
    error estimate, 131
    proof, 119
Rio Grande Gorge Bridge, 151
Ritalin, 42
Ritz, Walther, 176
Riverdale, 40
Robbins, Herbert, xvii, 12, 57, 179
Rochester, 102
Rockefeller Foundation, 19, 168, 178
rocket, 130, 131
Roman
    Catholic, 1
    Empire, 1
Romania, 1, 4
Romanians, 1
Roosevelt, President Franklin Delano, 11,
    12, 20, 28
Roosevelt, Theodore, 12
Rosenbergs, 25
Rosenbluth, Marshall, 17, 121
Rosenfeld, 12
Rothschild, Albert, 3
round-off, 53
    error, 53, 107
routine, 108
Runge, Carl, 175, 177
Runge, Nina, 177
Russell, Bertrand, 113
Russell, John Scott, 142, 143
Russia, 2, 12, 30
Russian, 92
Rusyns, 1
Rutgers, 181

Sakamoto, Reiko, 112
San Diego, 51
Sandia National Laboratory, 30
sanity, 109
Santa Fe, 20, 40, 159
    Railroad, 20
Santa Maria della Bomba, 27
Sarafite, Justice Joseph A., 75
Saratoga Springs, 45

Sarnak, Peter, 111
Saturday morning inspections, 21
Saul, Mark, 161
scalar product, 128
Scandinavian, 93
scattering, xvii, 133, 152
　data, 145, 151
　experiment, 151
　operator, 151, 153
Scattering Theory, 67, 82, 109, 150
scattering theory, 118
　for automorphic functions, 111
　in Euclidean space, 111
Scattering Theory for Automorphic Functions, 109, 150
Schauder bases, 108
Schauder, Julius, 108
scheme, 53
Scherrer, Paul, 177
Schnirelmann, Lev, 178
School of Engineering, 79
Schottky, Friedrich, 143
Schrödinger
　operator, 146
Schrödinger, Erwin, xvii, 58, 145, 169, 178
Schwartz, Jacob T. (Jack), 61, 62, 80, 163
Schwarz, Hermann Amandus, 98, 176
Schwarzschild, Karl, 175
scientific
　computation, 89, 121
　team, 33
Scientific American, 60, 85, 183
Seattle, 61
Second Avenue, 47
second derivative, 124, 131
Second Great War, 5
Second World War, 4, 178
second-order methods, 141
Secretary of War, 169
SED, 22
SEK, 6
Selberg, Atle, 100
Selected Papers, 109, 121
selection principle, 141
self-reference, 129
self-similar waves, 142
semi-group, 150, 153
Serbia, 1
Serre, Jean-Pierre, 94
Sethian, James, 61
seventieth birthday, 87
shallow water flow, 129
Shiffman, Max, 50, 180
shifting, 124
shock, 135, 137, 140
　capturing, 24, 32, 82, 133, 140, 141, 170
　flows, 32
　front, 137
　tracking, 140
　tube, 136, 137
　wave, 135, 141
　waves, xvii, 25, 81, 88, 112, 117–119, 124, 129, 133, 135–137, 140, 141, 169, 180
shockless airfoils, 119
shocks, 32, 112, 133–135, 137, 139, 140, 170
Shu, Chi-Wang, 61
SIAM, 86, 120
SIAM Review, 109
Sidrane, Kitty, 45
Sidranes, 45
Siegel, Carl Ludwig, 98–100, 178
signal, 124, 137
　propagation, 122, 123, 131
　strength, 123
signaling, 149
similar, 146
simple
　groups, 117
　waves, 140
simplifications, 114
sine, 127
　functions, 128
　waves, 127
sine-Gordon, 143, 147
sines, 128, 131
Singer, Isadore M., 94
Sistine Chapel, 58
six-dimensional phase space, 126
sixtieth birthday, 86
Skau, Christian, 114
Sloan Foundation, 20, 181
Slovaks, 1
Slovenia, 1
small differences, 131
Smoller, Joel, 64, 141
smooth
　flow, 117
　initial data, 126
　motions, 126
Snell, J. Laurie, 27
socialism, 4
socialist, 6
Society for Industrial and Applied
　Mathematics, 68, 82
solar system, 126
solid state, 150
solitary waves, 143, 145
soliton, 34, 145, 147, 148
solitons, 68, 88, 118, 133, 142, 143, 146, 149
solve systems of linear equations, 105
sonic boom, 135
Soros, George, 129
sound, 54, 122, 152
　barrier, 37, 135
　waves, 117, 152, 153
Southeast Asia, 70

Soviet, 67
  army, 4
  revolution, 114
Soviet Union, 29, 99, 181
Soviet-American joint mathematics, 65
space-time, 137, 140
spatial domain, 54
special solutions, 118
spectral
  representation, 154
  theory of automorphic functions, 154
speed, 125
  of sound, 135
Springer, Ferdinand, 177
springs, 126
St. Petersburg, 46
stability, 24, 53, 61, 133, 134, 170
  of a numerical scheme, 86
  theory, 117
stable, 53, 89
Stalin's Moscow, xvi
Stalin, Josef, 4, 5, 30, 85
Standard Model of Cosmology, 141
Stanford University, xvi, 17, 31, 36, 59, 86, 111, 150, 168
Star Wars, 27
state variables, 137
Staten Island University, 75
stationary points, 130
statistical mechanics, 112
statisticians, 60
steady state, 123
steel, 122
Steele Prizes, 91
Stein, Edith, 176
Still, Carl, 177
Stillwater, Kansas, 85
stochastic processes, 60
stock market
  crash, 4
  investing, 130
Stoker, James J., 49, 179, 181
Stokes, George, 114, 137, 143
Stokowski, Leopold, 46
Stoltenberg, Prime Minister Jens, 94
Stone, Marshall, 101, 113
Størmer, Carl, 93
storyteller, 25, 97
straight line, 129, 138
Strauss, Johann Jr., 2
Strauss, Lewis, 28
strong convergence, 139
Stubhaug, Arild, 94
Stuyvesant High, 17, 76
Super, 30
supercomputer centers, 86
supercritical wing sections, 112
superposition, 131

supersonic, 49
  flows, 180
Supersonic Flow and Shock Waves, 37, 136, 183
Supreme Court, 75
surface integrals, 137
Swarthmore College, 46
Sweden, 93
swim team, 39
Switzerland, 44
Sylow, Ludvig, 93
Synge, John, 37
Szálasi, Ferenc, 6
Szasz, Otto, 13
Szegö, Gábor, xvi, 8, 13, 17, 31, 168
Szegös, 31
Szilard, Leo, 9, 20, 24, 28, 30, 31
Szőkefalvi-Nagy, Bela, 8, 108
Sztójay, Döme, 6

Taiwan, 66
tangent, 129
  line, 130
Taos, 151
teaching, 88, 161
Tel Aviv, 35
Telefunken, 177
telegraph, 176
Telenor Group, 94
Teller, Edward, 11, 22, 24, 27, 30, 100, 170
temperature, 122, 131
Temple, Blake, 141
tennis, 17, 19, 39, 48, 65
Terrell, Maria, 82, 84, 112
Texas A&M, 19, 20, 36
The Book, 116
The Enjoyment of Mathematics, 11, 12
"The flowering of applied mathematics in America", 118
"The formation and decay of shock waves", 81, 135
The Mathematical Experience, 85
The Mathematics Journal for Secondary Schools, 8
The Science of Violin Playing, 46
The Voice of the Martians, 9
theorem
  closed graph, 53
  divergence, 137
  Lax-Richtmyer equivalence, 133
  Schauder fixed point, 108
  Stokes', 137
theoretical, 89
theory
  algebraic number, 100, 113
  analytic number, 109
  functional, 153
  Lax-Phillips scattering, 109

linear operator, 152
logic and set, 88
manifold, 55
measure, 55
particle, 54
quantum field, 148
scattering, 28, 61, 86, 88, 111, 150–153
spectral, 178
thermal convection, 63, 64
thermalize, 145
thermonuclear fusion, 79
Thomas, Elizabeth, 40
Thompson, Gerald L., 27
thunder, 135
Toda, 143
  lattice, 147–149
Toda, Morikazu, 148
Toeplitz, Otto, 11, 175
Toeplitz/Caratheodory, 121
topologists, 80
topology, 110
  point-set, 55
total variation
  diminishing, 82
traditional mathematics, 116
trajectory, 125, 129, 130
Transcendental Students, 75
transfinite diameter, 168
transform
  Fourier, 111, 121
  Radon, 111
transient, 123
  distribution, 123
translation representation, 154
transonic flows, 37
Transylvania, 4
Treaty of Trianon, 4
triangle, 154
Trinity, 28
Trivisa, Konstantina, 68
Trondheim, 95
Truesdell, Clifford, 91
Truman, President Harry, 29
Tulane University, 13, 92
Turan, Paul, 8
turbulence, 62
Turing machine, 117
Turing's definition of calculation, 117
Turkish
  Empire, 1
  government, 169
Twain, Mark, 33
two imaginary points at infinity, 99
two-dimensional canvas, 114
two-slit experiment, 54

U.S., 4
  Army, 11, 30
  Constitution, 69
  military, 20
  National, 92
  Naval Academy, 81
  visas, 12
ugly theorems, 116
Ukraine, 55
Ukrainian Academy of Sciences, 98
Ukrainians, 98
Ulam, Francoise, 159
Ulam, Stanisław, 22, 24, 27, 30, 144, 159, 168
Un-American Activities, xvi, 56, 85
uncle Imre, 15
undergraduate
  curriculum, 83
  education, 83
underground testing, 30
uniform convergence, 84
Union Canal, 142
uniqueness, 117
unitary group, 153
United Nations, 30
UNIVAC, 51, 61, 181
UNIVAC I, 52
Universidade de Rio de Janeiro, 87
University
  Archive, 77
  of Birmingham, 98
  of Breslau, 175
  of California, 51
  of California at Berkeley, 62
  of Chicago, 97, 101
  of Edinburgh, 94
  of Göttingen, 60
  of Michigan, 55, 64
  of New Mexico, 30, 31, 36, 60, 71, 78, 87, 111, 159
  of Oslo's Department of Mathematics, 94
  of Rochester, 102
  of Texas, 20
  of Tokyo Press, 82
  of Zürich, 175
  Senate, 69
unstable schemes, 53
Upper West Side, 19, 65
uranium
  235, 28
  bomb, 29
  fission, 9

V-8, xx
Varadhan, S. R. Srinivasa, 94
variable
  coefficients, 111
  real, 55, 63
Veblen, Oswald, 19, 179
vector spaces, 108

vectors, 126, 128
velocity, 130, 131
Venice, 87
Veronka, 39, 87
vertical asymptote, 130
Vészi, 6
vibrating system, 107
vibrations of continuum mechanical systems, 178
Vienna, 1–3
Vienna Awards, 5
Vietnam, xv, 77
  War, 69
Vietnam-Cambodian War, 78
Villa La Pietra, 87
Village Fighting, 20
Vinnikov, Victor, 157
viola, 46
violets, 3
violin, 46
Virgin of Guadalupe, 113
visualization, 126
volume integrals, 137
von Beethoven, Ludwig, 1
von Braun, Werner, 97
von Kármán Prize, Theodore, 120
von Kármán, Theodore, 3, 11, 97, 168, 175
von Neumann
  algebras, 169
  criterion, 24, 133, 134
  John, xv, xvi, 2, 3, 8, 13, 17, 19, 22, 24, 27, 29–31, 33, 86, 88, 100, 101, 116, 128, 133, 134, 137, 139, 140, 144, 145, 160, 167, 178
  Lecture, 68
vortex blob method, 62

WAC, 25
Wald, George, 77
Walker, Homer, 64
war, 76, 169
Warren Weaver Hall, 72, 74–76
Washington, 18, 51, 76
  D.C., 71
  Irving High School, 46
  Square College, 79
  Square Park, 50
water, 142, 148
  polo, 5
  waves, 32, 142, 143, 148
Waterloo Bridge, 7
Watson, George Neville, 98
wave
  information, 141
  motion, 54
  nature, 54
  of translation, 143
  problem, 140

propagation, 64, 117, 152
Wave of Translation, 143
wavelets, 54
weak
  convergence, 108, 139
  sense, 141
  shocks, 140
  solution, 141
weather, 130
  prediction, 130
Weaver, Warren, 50, 180
Wehrmacht, 14
Weierstrass, Karl, 88, 98, 177
weight functions, 66
Weil, André, 100, 157, 178
Weingarten, Julius, 110
Weinstein, Sol, 20
Weisz, 3, 8
Weld, Tuesday, 19
well-posed initial boundary value problem, 122
Wendroff, Burt, 60, 94, 111, 133, 159
West
  90th, 19
  90th Street, 19
  91st, 19
  Side, 19
Weyl, Hermann, 100, 102, 168, 169, 177, 178
What Is Mathematics?, xvi, 57, 179, 183
wheelchair, 39
White Terror, 4
Wiener, Norbert, 178
Wigner, Eugene, xvii, 24
Wise, Stephen S., 19
Wolf Prize, 99
Wolf Prize of Israel, 86
Wolf, Dr. Ricardo Subirana y Lobo, 86
Wolf, Francisca, 86
Wolfe, Dr. Robert, 74, 77
Wolff, Lucile Gardner, 182
Wollman, Steve, 61
World War I, 42, 99
World War II, 5, 49, 135
Wouk, Herman, 19

X-ray crystallography, 121

yellow fever, 29
Yellow Peril, 177
Yiddish, 65
Yugoslavia, 4

Zabusky, Norman, 143, 145
Zalcman, Lawrence, 88, 112
Zarca, Bernard, 113
Zermelo, Ernst, 175
Zionist, 6
Zsigmund, 3

Zürich, 49, 168, 179
Zwas, Gideon, 82
Zygmund, Antoni, 54, 97, 111